THE JUNKERS Ju 52 STORY

JAN FORSGREN

FONTHILL

Fonthill Media Language Policy

Fonthill Media publishes in the international English language market. One language edition is published worldwide. As there are minor differences in spelling and presentation, especially with regard to American English and British English, a policy is necessary to define which form of English to use. The Fonthill Policy is to use the form of English native to the author. Jan Forsgren was born and educated in Umeå; therefore British English has been adopted in this publication.

Fonthill Media Limited
Fonthill Media LLC
www.fonthillmedia.com
office@fonthillmedia.com

First published in the United Kingdom and the United States of America 2016

British Library Cataloguing in Publication Data:
A catalogue record for this book is available from the British Library

ISBN 978-1-78155-515-6

Typeset in 10.5pt on 13pt Minion
Printed and bound in England

Author's Note

Much has been written about the Junkers Ju 52/3m. This is not the first book to cover this particular type of aeroplane, nor will it be the last. The purpose of the book is to provide a general history and overview of the Ju 52/3m, from its first flight in 1930 as a single-engined aeroplane intended for small freight carriers until the present day as an aviation museum exhibit. The book begins with an introduction to Professor Hugo Junkers and the early days of the Junkers Flugzeug-und Motorenwerke AG and its predecessors, as well as the various civilian airliners developed by Junkers before 1930.

This is followed by a chapter on the development of the Ju 52 and the three-engined Ju 52/3m. A technical description of both the basic Ju 52 and Ju 52/3m includes brief details of variants. Production in Germany as well as France, Hungary, and Spain are covered in the next chapter.

The high-profile Lufthansa and Luftwaffe use of the Ju 52/3m are then covered in separate chapters. Regarding the latter, I have elected to use the German word *Transportverbände* (literally transport units) throughout the book. Other Luftwaffe vocabulary such as *Staffel* (Squadron), *Gruppe* (Group), and *Geschwader* (Wing) are also used in the book instead of their British equivalents. With the Ju 52/3m seeing use both by various airline companies and air arms in nearly fifty countries (from Argentina to Yugoslavia), these aircraft are covered in the following two chapters. Included here are the Ju 52/3ms built in France and Spain as the AAC.1 and CASA 352 respectively. With little information on these aircraft having appeared in print previously (in English-language publications at least), fairly extensive chapters on their respective use have been included. The final chapter contains information on preserved Ju 52/3ms, including both complete aircraft and major components.

Extensive appendices include Lufthansa Ju 52/3ms, Luftwaffe Ju 52/3m units, and AAC.1 and CASA 352 production lists. A bibliography is also included. An index covering names, geographical locations, and types of aircraft can be found at the end of the book.

Acknowledgements

The process of writing usually means long, lonely hours before the book has been finished. The following individuals, listed in alphabetical order, have all contributed to the content of the book: eminent Italian aviation historian and author Gregory Alegi provided photos and information about the Ju 52/3m in Italian service; John Austin-Williams (JAWS), for much information and contacts on the CASA 352 in South Africa; Mats Averkvist, for allowing me to use photos of Ju 52/3ms in Norway; Andrew Bird, for the photo of the CASA 352 preserved at the Aviodrome; Hans Boëthius, for information on SAS Ju 52/3ms; Belgian aviation historian Daniel Bracx, for valuable contacts; Paul Branke, for photos and information on Estonian and Luftwaffe Ju 52/3ms; Winston Brent, for providing lots of valuable contacts regarding South African Ju 52/3ms; Jean-Pierre Dubois, Air-Britain's expert on the *Aéronautique navale*, for much assistance regarding French Ju 52/3ms; my old friend Pär Erixon, for allowing the use of his photos of a Ju 52/3m wreck taken in northern Norway in 1980; Athol Franz, for information about the CASA 352 currently flying in South Africa; Spanish aviation historian and expert on the Ju 52/3m and CASA 352 in Spain, Luis González Pavón, who immediately made available a number of photos as well as much information; Albert Grandolini, for supplying a vast amount of information and photos on the AAC.1 in French military service; Robert S. Grant, for allowing me to use his photos of CASA 352 CG-ARM; the President of FAI John Grubbström, for locating several contemporary photos of Junkers aeroplanes in the FAI Archives; Sonja Grünbauer, for making available her wonderful photos of the CASA 352 still flying in South Africa; Dan Hagedorn, for assistance on the Ju 52/3m, and particularly the ubiquitous Aconcagua, in South and Latin America; Bengt Hermansson of the Forced Landing Collection, for providing photos and information on the CASA 352 and the Ju 52/3m g14e cockpit mock-up preserved at Ugglarp; my friend Ulrich Hild, for photos of preserved Ju 52/3ms in Germany; Pekka Karhumäki, for much information on Aero O/Y Ju 52/3ms; Michael Magnusson, who most kindly provided a wealth of photos and information on the Ju 52/3m in Argentina and Uruguay; Nils Mathisrud, for information on the Ju 52/3m in Norway; Dave McDonald, for his considerable help in locating photos of Australian-registered Ju 52/3ms; Yolanda Meyer of the Transnet Heritage Library, for finding several rare images of DETA and South African Airways Ju 52/3ms; Canadian author and aviation historian Larry Milberry, for forwarding a multitude of photos and information on the Ju 52 in Canada; David Nicolle, for information on AAC.1s in the Middle East; Mikael Olrog, for information on preserved aircraft around the world; my friend Miro Patocka, for diligently searching for information about the Ju 52/3m in Czech and Slovak service, as well as graciously loaning a number of books; Thijs Postma, for allowing the use of Ju 52/3m photographs from his own collection; Michael Sanz, for locating suitable photos in the Swedish Aviation Historical Society's Archives; retired Kar-Air airline Captain Pertti Seeve, for invaluable assistance regarding Spanish and Finnish Ju 52/3ms; Brett Stolle of the National Museum of the USAF, for making a number of photographs of Luftwaffe Ju 52/3ms available; Björn Svedfeldt, for information on the acquisition of CASA 352 T.2B-142 currently preserved at Björn's air museum at Ugglarp; Italian aviation historian Paolo Varriale, for several valuable contacts; Detlef Walther, for much contemporary documentation on the Ju 52/3m; aviation writer and historian John Wegg, for valuable contacts; Danish aviation historian Tom Weihe, for providing information on Danish-registered Ju 52/3ms; Luc Wittemans, for much information on Belgian-operated Ju 52/3ms; and last, but certainly not least, Björn Zereba, for the photo of the Ju 52/3m preserved in Berlin.

A special thank you to Bernd Huckenbeck of Ju Air. Also thanks to my editor at Fonthill Media, Jay Slater, for having faith in the project. Finally, thanks to Jessica for her incredible lack of interest in old aeroplanes, as well as lots of laughter.

Apologies to anyone that I have forgotten.

Contents

Notes on Sources

Ron Davis's book *Lufthansa* gave brief details on the Ju 52/3m in service with the German national airline, both at home and abroad. Information on Lufthansa's individual Ju 52/3m was culled from several sources, including the aforementioned *Lufthansa*, but also the website www.airhistory.org.uk, which contained civilian aircraft registers until 1939. Regarding aircraft losses, aviation-safety.net has been referred to. Contemporary magazine articles, such as *Shell Aviation News* and *Aeroplane*, provided some interesting stories, particularly regarding the rescue flight to Kabul in September 1937. Regarding the Ju 52/3m in Luftwaffe service, Martin Pegg's two volumes on *Luftwaffe Transport Units 1937–1945* provided a wealth of information. Information on Luftwaffe Ju 52/3m transport units was, to a large extent, compiled from Michael Holm's excellent website www.ww2.dk.

Various books by Chris Shores *et al.*, including *Fledgling Eagles*, *Air War for Yugoslavia, Greece and Crete*, and *A History of the Mediterranean Air War*, proved invaluable in establishing a day-to-day account of Luftwaffe Ju 52/3m operations. The use of Ju 52/3ms in service around the world was compiled from a wide variety of primary archival and second-hand published accounts, both in print and available on the internet. The AAC.1 and CASA 352 Production Lists could not have been compiled without the gracious assistance by Jean-Pierre Dubois and Luis Gonzalez Pavón.

List of Abbreviations

ABA	*Aktiebolaget Aerotransport* (Swedish Airlines)
ARR	*Aeronautica Regalâ Românâ* (Royal Rumanian Aeronautics)
CASA	*Construcciones Aeronáuticas Sociedad Anónima* (Spanish Aircraft Manufacturer)
CAUSA	*Compañia Aeronáutica Uruguaya SA* (Uruguay Aeronautics Company Ltd)
C/n	Construction number
ČSA	*Československé Státní Aerolinie* (Czech Airlines)
CVL	*Československé Vojenské Letectvo* (Czech Air Force)
DDL	*Det Danske Luftfartsselskab* (Danish Airlines)
DLH	*Deutsche Luft Hansa* (name changed to Deutsche Lufthansa A.G. in 1933)
DNL	*Det Norske Luftfartselskap AS* (Norwegian Airlines)
EASSU	Enemy Aircraft Servicing and Storage Unit
EEES	*Elliniki Etairia Enaerion Synhinoion AE* (Greek Airlines)
EVA	*Elliniki Vasiliki Aeroporía* (Royal Hellenic Air Force)
JKRV	Royal Yugoslav Air Force (1918–1941)
JRV	*Jugoslovensko Ratno Vazduhoplovstvo* (Yugoslav Air Force, 1945–1992)
LARES	*Liniile Aeriene Române Exploatate de Stat* (Romanian State-Operated Airlines)
NII VVS	*Nauchno-Issledovatel'skiy Institut Voyenno-Vozdooshnykh Seel* (Scientific Research Aero Institute Military Air Forces)
Ntu	Not taken up
ÖLAG	*Österreichische Luftverkehrs AG* (Austrian Airlines AG)
RAF	Royal Air Force
RHAF	*Magyar Királyi Honvéd Légierö* (Royal Hungarian Air Force)
SAAF	South African Air Force
SABENA	*Société Anonyme Belge d'Exploitation de la Navigation Aérienne* (Belgian Corporation for Air Navigation Services)
S/GMMTA	*Sous-Groupement Moyens Militaires Transport Aériens—Armée de l'Air* Sub-Group of Military Vehicles Air Transportation—French Air Force
SVZ	*Slovenské Vzdušné Zbrane* (Slovak Air Force)
T.Col	*Tenente Colonello* (Lieutenant Colonel, *Regia Aeronautica*)
W Nr	*Werk Nummer* (Construction Number)

A Luftwaffe Ju 52/3m seen taxiing on a primitive airfield on the Eastern Front. (*NMUSAF*)

Introduction

Along with the Douglas DC-3, the Junkers Ju 52/3m ranks among the best-known transport aircraft of all time. Instantly recognisable, the Ju 52/3m was used by numerous airlines and military air arms, performing the mundane, but essential task of carrying passengers and freight. For many the Ju 52/3m is forever linked with the Luftwaffe. Here the Ju 52/3m played a crucial role in many operations during the Second World War. Even prior to the outbreak of war in September 1939, Ju 52/3ms of the Legion Condor made international headlines when the small Basque town of Guernica was bombed. German parachutists were dropped into combat over Norway, the Netherlands, and Crete. Airlift operations at Demyansk, Stalingrad, and North Africa depended on the Ju 52/3m. For others, it was a really modern European airliner used in quantity by airlines in Europe, Southern Africa, China, and South America, helping to build a network of international air routes between the continents. After the end of the Second World War, the Ju 52/3m entered production in France and Spain as the AAC.1 and CASA 352. Today several aircraft survive in aviation museums, with a few still flying.

First flown in 1930, the initial Ju 52 was powered by a single engine. Originally designed with the aim of retaining the Junkers Company's foothold in the burgeoning air freight business, the Ju 52 was a cargo monoplane. It was estimated that the air freight sector would increase substantially during the early 1930s, making it a sure-fire success for future investment. Having a cargo compartment of 16.7m^3 and being of rugged all-metal construction, the Ju 52 had, it was thought, a good chance of becoming the mainstay of air freight companies around the world. Apart from the above characteristics, it was also designed to be cheap and easy to operate. Passenger transport was, at the time, deemed to be of less importance. In the end, the air freight business did not develop as originally thought, with Junkers abandoning its original intention of establishing an air freight company of its own.

Instead, transporting passengers by air became a growing part of the civilian airline business. Following early sales to German operators, as well as Canada, production of the single-engined Junkers Ju 52 was abandoned. The basic airframe was developed into a three-engined airliner, being designated as the Ju 52/3m, the '3m' indicating three engines. In the event, the Ju 52/3m was to become the largest passenger airliner in Europe, serving with airlines in thirteen countries. With the German airline Lufthansa having interests in South American and Chinese airlines, the Ju 52/3m became a mainstay of airline operations in these parts of the world. In 1932, Bolivian Ju 52/3ms were used to carry troops and supplies during the Gran Chaco War. Additionally, South African Airways and DETA in the Portuguese colony of Mozambique ordered Ju 52/3ms.

Even though some 400 Ju 52/3ms entered service with civilian airlines before 1939, the Ju 52/3m is forever linked with the Luftwaffe. Initially, it was selected as an auxiliary bomber, equipping bomber units of the then still-secret Luftwaffe. With the introduction into service of more modern bomber aircraft, the Luftwaffe relegated the Ju 52/3m to transport duties.

Seeing its European baptism of fire during the Spanish Civil War, Ju 52/3ms bombed the Basque town of Guernica. It was used extensively on every front during the Second World War, from the Arctic to North Africa, from the Blitzkrieg in the West to Stalingrad in the East—the Ju 52/3m was used for a multitude of missions. Even though it was thoroughly obsolete by the start of the war, production levels increased substantially. It was loved by crews and troops alike, earning several nicknames, including '*Tante Ju*' ('Auntie Ju') and '*Eisernes Annie*' ('Iron Annie'). The Ju 52/3m can be described as rugged, dependable, and decidedly utilitarian. One German pilot described it as virtually 'indestructible'; however, the ruggedness of the Ju 52/3m was not matched by its looks. It had, in the words of legendary British test pilot Eric 'Winkle' Brown, the appearance of a '… fantastic amalgam of corrugated skinning, trailing flappery and heavily-braced undercarriage, not to mention a trio of air-cooled radial engines, which, as I was to discover, possessed all the discordance of a phalanx of over-revving lawnmowers [and] demonstrated a total disregard for the most elementary aerodynamic considerations!'[1]

The Ju 52/3m was the most numerous transport aircraft to serve with the Luftwaffe, similar to the widespread use of the Douglas DC-3/C-47/R4D/Dakota by the Allies. Compared with the Douglas transport, the Ju 52/3m proved to be rather more adaptable, seeing use as a bomber, paratrooper carrier, towing cargo gliders, cargo transport, general and VIP transport, and advanced trainer. The trimotor was also used in some numbers in the dangerous task of degaussing magnetic mines. It was also used as an engine test bed, testing both a variety of air-cooled and liquid-cooled engines. Although the basic Ju 52/3m airframe remained essentially the same, it was continually refined. It was easy to fly and maintain, albeit slower and carrying a smaller payload than the larger DC-3. Other advantages were the fact that the Ju 52/3m could operate from rough, primitive strips that simply would not accommodate the DC-3. The Ju 52/3m could be fitted with floats (tested on the C-47, but not adopted for service) or skis in place of a conventional wheel undercarriage, thus making it possible to operate from different kinds of surfaces in all kinds of weather.

Before September 1939, about 2,000 Ju 52/3ms had been delivered, with the Luftwaffe receiving the lion's share. Large quantities of Ju 52/3ms were lost during the war, particularly during the invasion of the Netherlands in May 1940, the invasion of Crete in May 1941, the evacuation of Axis troops from North Africa in April and May 1943, and on the Eastern Front, and then particularly during the Stalingrad airlift between November 1942 and January 1943. In fact, the losses in aircraft and aircrew were so great that the *Transportverbände* of the Luftwaffe were never able to fully recover.

Between September 1939 and September 1944, 2,804 Ju 52/3ms were built, with the rugged, but thoroughly obsolete transport being phased out of production. With German forces retreating on every front, priority was given to producing combat aeroplanes. Although exact production quantities are not known for certain, it is believed that nearly 5,000 Ju 52/3ms were built. Although the vast majority of Ju 52/3ms were built in Germany, production lines were also established during the war in France and Hungary. In spite of improved and more capable transport aircraft being introduced into the Luftwaffe service from 1944, particularly the Ju 252 and Ju 352, these aircraft never replaced the venerable Ju 52/3m. Apart from Lufthansa and Luftwaffe, the Ju 52/3m was operated by over thirty military air arms, as well as civilian companies in more than forty countries.

Amazingly, a few captured Ju 52/3ms were rebuilt post-war, seeing brief use as airliners in Britain. The basic soundness of the design (and the availability of a production line along with jigs and spares support) saw the continuation of Ju 52/3m production in France as the AAC.1. No less than 415 AAC.1s were built, both for the French armed forces and airlines such as Air France. The AAC.1s of the *Armée de l'Air* were heavily used in Indochina, both as transports and as *ad hoc* bombers, as well as in Algeria. Production was also initiated in Spain, with 170 being built between 1945 and 1954 as the CASA 352. Only operated by the *Ejército del Aire* (Spanish Air Force, EdA), the final CASA 352s were not withdrawn until the late 1970s.

The longevity of the Ju 52/3m can be seen from the fact that the last operational civilian aircraft was withdrawn from service in the early 1960s, with the trio of Ju 52/3ms originally delivered in 1939 to the Swiss Air Force lasting until 1982. Around fifty Ju 52/3ms survive, including post-war CASA 352s and AAC.1 derivatives, with eight being airworthy.

The Ju 52/3m is indeed one of the most classic of all aeroplanes, and its place in aviation history remains secure.

1

The Junkers Flugzeug-und Motorenwerke AG and its Predecessors

The German aircraft designer Hugo Junkers was one of the foremost pioneers of aircraft construction, particularly with regards to the use of stressed-skin metal airframe structures. Between 1915 and 1945, the aircraft manufacturer Junkers in its various constellations developed and built a wide variety of all-metal aeroplanes, ranging from fighters to huge passenger transports, more often than not powered by Junkers-designed engines.

Among its designs, the F.13 airliner, Ju 52/3m airliner and military transport, Ju 87 dive-bomber (better known as the Stuka), and the Ju 88 multi-role combat aircraft are perhaps the most well-known. During the latter stages of the war, the radical four-engined Ju 287 jet bomber was flown. It never entered service, remaining a prototype. In fact, only one Junkers design was constructed out of wood, this being the Ju 322 transport glider. Developed at the request of the *Reichsluftfahrtministerium* (RLM), the Ju 322 was a complete failure, partly due to Junkers' lack of experience in wooden aircraft design. Following the end of the war, the Junkers company was disbanded. Since the reformation of the German aviation industry, Junkers is part of Messerschmitt-Bölkow-Blohm.

The German aviation pioneer Hugo Junkers was born on 3 February 1859 at Rheydt, near Düsseldorf. His father was the owner of a textile factory. Junkers graduated from the Berlin Technical High School, also studying at the Karlsruhe and Aachen Technical Colleges. In collaboration with Wilhelm Oechelhäuser, a two-stroke piston engine was tested in 1892. Three years later, Junkers founded the Junkers und Compagnie, which specialised in manufacturing gas-fired boilers as well as cooling equipment. In 1897, Junkers received a professorship at the Aachen Technical College. A second company, the *Kaloriferwerk Hugo Junkers* was established at Dessau, producing air-heating gear and radiators. Revenues were used to finance various research projects.

The first decade of the twentieth century marked the initial flight of the heavier-than-air aeroplane. The practical applications of the aeroplane, such as carrying mail and even passengers and for military purposes, were soon to be recognised. The vast majority of the aeroplanes designed and built during this time were constructed out of wood and fabric, with only a few components being built out of various metals, including the engines. The capabilities of these early aeroplanes were limited. Professor Junkers' interest in aeroplane design began in earnest following a meeting with Professor Hans J. Reissner. Since 1907 Reissner had been involved in the development of a monoplane fitted with wings constructed out of duraluminium. The solution that Junkers arrived at was to attempt to eliminate parasite drag. In the event, Junkers filed a patent on 1 February 1912 for a thick cantilever aerofoil to be built entirely out of aluminium. The aerofoil body included the engine as well as room for pilot and passengers.

Already, Junkers was thinking about transporting passengers and cargo by air. In 1914, Junkers constructed a wind tunnel at Frankenberg (near Aachen); during the following five years, he used this wind tunnel to perform around 4,000 aerodynamic tests regarding the relationship between the shape of aerofoils, camber, and thickness. In Junkers' own words: 'What I wanted to achieve here, as well in my other work, was a drive into an unknown land, a pioneering task… such a task was the all-metal aircraft with strut-free wings.'[1]

The outbreak of the First World War in August 1914 saw a rapid increase in the development of the aeroplane for war purposes. Encouraged by his wind-tunnel tests at Frankenberg, Junkers attempted (on 2 February 1915) to get the Imperial German War Ministry to fund the construction of an all-metal aeroplane. Although the War Ministry initially turned down Junkers' request, funding was allocated in July 1915 after Junkers had delved further into the metal-forming process. The result was the Junkers J 1, which entered ground trials shortly before Christmas, eventually making its maiden flight on 18 January 1916, with *Leutnant* von Mallinkrodt at the controls. A single-seat cantilever monoplane powered by a 120-hp Daimler inline engine, the Junkers J 1 showed the way to the future regarding aircraft construction. It was unofficially referred to as the '*Blechesel*' ('Tin Donkey'). However, both the Junkers J 1 and its subsequent development, the J 2, were seriously overweight, reducing both rate of climb and overall performance.

An agreement was eventually reached with Anthony 'Tony' Fokker regarding the formation of a new company, the *Junkers-Fokker Werke AG Metallflugzeugbau* (established on 20 October 1917), which would develop and build all-metal combat aeroplanes.

The next Junkers design to be built was the Junkers J 4. A large two-seat biplane, it entered service as the Junkers J.1 in the summer of 1917. Featuring heavy armour protection, the J.1 saw fairly large-scale use in the ground-attack role, with more than 200 being built. Further refinement of the basic J 1 and J 2 design resulted in the Junkers J 7 single-seat fighter. Comparative trials, during which the J 7 was test flown by a number of front-line fighter pilots, including Manfred von Richthofen, showed the Junkers fighter to have a markedly better overall performance than other German fighters. In spite of this, the Idflieg fighter aircraft representative made a despairing remark about the Junkers fighter to the effect that the 'all-metal aircraft hullabaloo [being] a fashionable sickness'.[2] On his part, Professor Junkers was adamant that wooden aeroplanes would eventually become obsolete, and that all-metal aircraft represented the only way forward in order to obtain more capable aeroplanes.

In the event, the J 7 and its J 9 development entered limited production as the Junkers D.I, with forty being completed by early 1919. Although the Junkers D.I did not see service on the Western Front, several D.Is and the two-seat CL.I variant saw action against Bolshevik forces in the Baltic region in the immediate post-war era.

2

A Shift to Civilian Transport Aeroplanes

Following the end of the First World War, the market for warplanes disappeared overnight. Cooperation with Fokker ended, with the company being reformed on 24 April 1919 as the *Junkers Flugzeugwerke* AG. Instead of building fighters and bombers, Junkers turned his attention towards civilian aircraft, and, more specifically, the passenger airliner. The first wholly civilian aeroplane was the Junkers J 13 (later redesignated as the F.13), an open-cockpit monoplane that could carry four passengers in an enclosed cabin. First flown at Dessau on 25 June 1919, the Junkers F.13 proved to be a winner. It was both cheap to operate and of rugged construction. Almost 350 were built, with many airlines, including Lufthansa, Belgian SABENA, ABA (Swedish Airlines), and the Polish LOT operating the F.13 with considerable success. At the time, British and French aircraft manufacturers were delivering *ad hoc* airliners converted from bombers, such as the Airco DH.4 and Breguet 14, the Junkers F.13 was a radical step forward. It can be said that the F.13 provided the foundation and first stepping stone of the modern airline industry. The final Junkers F.13s were withdrawn from use in the late 1940s, with a few of these very important and ground-breaking aeroplane being preserved in aviation museums.

With the ratification of the Treaty of Versailles in January 1920, Germany was forbidden to produce both aeroplanes for both military and civilian purposes. The ban would last for six months for civilian aeroplanes. The Inter-Allied Aeronautical Commission of Control (IAACC) was able to confiscate German aeroplanes, both old ones and newly built Junkers F.13s. The wartime combat aeroplanes were invariably destroyed, with a quota of every fourth Junkers F.13 completed to be handed over to the IAACC.

In 1921, the Junkers formed an affiliate company, the Junkers Luftverkehr AG, which was to promote the Junkers F.13 to prospective customers, as well as initiate airline operations in various parts of the world, including Persia (present-day Iran), China, and South America. Junkers Luftverkehr would also operate a large number of routes in Germany and other parts of Europe. In fact, Junkers Luftverkehr was an umbrella organisation that included some fourteen different airlines and one flying school. Junkers had large shares of stocks in each of these companies.

During the early 1920s, factories were established at Limhamn, Sweden, and Fili, USSR, with aircraft of various Junkers designs being produced both for domestic operators and export customers. The Fili factory was handed over to Soviet authorities in 1927, eventually being redesignated as GAZ No. 7. However, German and Soviet cooperation regarding the development of all-metal aircraft for military purposes continued, as related below.[1]

A separate affiliate engine company, the *Junkers Motorenbau GmbH* was formed in 1923. Although remaining in production until 1932, the single-engined Junkers F.13 was soon developed into the slightly larger and more capable Junkers W 33. Powered by a 305-hp Junkers L5 inline engine,

The world's first modern airliner, the Junkers F.13. (*John Grubbström/FAI Archives*)

The single-engined Junkers W 33/34 series was developed from the Junkers F.13. (*John Grubbström/FAI Archives*)

the W 33 was first flown on 17 June 1926. The W 33 soon established itself as a rugged and dependable aeroplane, being used on a number of spectacular long-distance flights. On 13 April 1928, one W 33 named *Bremen* became the first aeroplane to cross the Atlantic in an east–west direction. Produced in limited numbers, the W 33 was subsequently modified with a radial engine, being redesignated as the Junkers W 34. Large quantities of W 34s were built, with many being used by the Luftwaffe as a light transport and advanced trainer. A further development was the Junkers Ju 46, which essentially was a W 34 with a strengthened airframe. The Ju 46 was used by Deutsche Lufthansa as an airmail carrier, being catapulted off ships to deliver the mail before the ship entered port.

A larger, three-engined airliner, designated as the Junkers G 24, appeared in 1924. Carrying a maximum of nine passengers, the G 24 was a definite improvement over the earlier F.13. With the Treaty of Versailles placing restrictions on engine power, payload, and performance, a lower-powered model of the G 24 was marketed as the G 23. In all, eighty-five G 23s and G 24s were built, operating with a large number of airlines and air arms.

In September 1926, another trimotor Junkers airline flew for the first time. Designated by Ernst Zindel, the Junkers G 31 was promoted as 'The airliner for long distance traffic'. Apart from the prototype, only twelve Junkers G 31s were built, with one (c/n 3003) being the thousandth aircraft produced by Junkers. Even though only thirteen Junkers G 31s were built, the G 31 provided an important stepping stone in the evolution of the Ju 52/3m.

For its time, the G 31 was a successful airliner; however, it is today perhaps best remembered hauling gold in New Guinea. From 1931 until early 1942, Guinea Airways operated four Junkers G 31s. Following the Japanese invasion, two G 31s were pressed into service with the Royal Australian Air Force. Both the G 23/24 and G 31 showed the way for things to come, having the same basic configuration as the later Junkers Ju 52/3m. In this context, it can be said that the trimotor configuration for airliners became more or less standard during the late 1920s. The American Ford company produced the Ford Trimotor, while the Dutch Fokker F.VII/3m, French Wibault 280T12 series, and Italian Savoia-Marchetti SM 73 and 83 all shared the airways with the Ju 52/3m. However, it was the Ju 52/3m that would become the most widespread and widely known of all the different three-engined airliners, being produced in far greater numbers than all the others combined.

During the late 1920s, Junkers' fascination with the *Nurflügel* (all-wing aeroplane) construction began of what would eventually result in the Junkers G 38, which first flew in 1932. Of enormous

The G 24 was the first three-engined Junkers airliner. (*John Grubbström/FAI Archives*)

Although the Junkers G 31 was built in limited numbers, it can be considered as the immediate predecessor of the Ju 52. (*Arlanda Civil Aviation Collection*)

At the time, the massive G 38 was the largest airliner built by Junkers. (*John Grubbström, FAI Archives*)

size for its day, only two G 38s were built. One crashed in 1936, with the second being mostly used for demonstration flights and development work.[2] While the outsized G 38 could be considered a bold step into the future, a more modest transport aeroplane was also under consideration.

In late 1925, a Junkers design team, headed by Kurt Weil and Hans Bongers, initiated work for a cargo aeroplane intended for the Junkers Luftverkehr company. This may be seen as the genesis of the future Ju 52.

However, the actual foundation for what would evolve into the Ju 52 can be traced to one of Junkers managers, Gotthard Sachenberg, who specified that a new transport aeroplane should be able to carry a payload of 2.2 tons over a range of 800 km (497.1 miles). The new aeroplane was to be specifically aimed at the bourgeoning air cargo market, and, apart from having a large payload, was to be of rugged and simple construction, with low operating costs. In more ways than one, what was envisaged was a flying truck that could operate from short, primitive airstrips. Design work on the new cargo aeroplane (designated Junkers Ju 52) began in 1929.

The designer of the Ju 52 was *Diplom-Ingenieur* (Certified Engineer) Ernst Zindel, who in 1927 had succeeded Karl Plauth as Junkers' chief designer following the latter's death in an air accident. Zindel was born in Mistelbach on 23 January 1897. When war broke out, he volunteered for service in the German Imperial Army. After suffering severe injuries, Zindel signed up at the Charlottenburg Technical College, studying shipbuilding. In 1920, he joined the Junkers company. Following the death of Otto Reuter (Junkers' chief designer) in 1922, Zindel was appointed to work at the research department, which was headed by Professor Otto Marder.

Zindel was responsible for the design of the Junkers G 24 and G 31 trimotors, as well as the single-engined W 33/34 light transport aircraft. In 1927, Zindel became the head of the project development department. Five years later, he became responsible for design, development, and flight testing of all Junkers designs, including the huge G 38 and Ju 90 airliners, the Ju 88 and Ju 388 medium bombers, and the Ju 290 and Ju 390 maritime reconnaissance aircraft. However, the aircraft that Zindel would forever be associated with is the Junkers Ju 52/3m.

Following the end of the war, Zindel worked for the P.I.V. Company at Bad Homburg vor der Höhe, where he designed the infinitely variable gearbox. Ernst Zindel died at Bad Homburg vor der Höhe on 10 October 1978, aged eighty-one.

The Ju 52 prototype, registered as D-1974, made its maiden flight on 11 September 1930, with Junkers test pilot Wilhelm Zimmerman at the helm.[3] The Ju 52 prototype was initially powered by one 600-hp BMW VIIaU inline engine. It had been planned that the Junkers' company's own L88 engine would be fitted to subsequent production Ju 52s. However, it was not until 3 September 1931 that D-1974 was air tested with the L88, with the four-bladed Heine wooden propeller needed to drive the Junkers L88 having a diameter of 5 metres. In the meantime, the basic Ju 52 airframe had been slightly modified, with the dihedral of the wings being increased, along with a slight sweepback in order to compensate for the shifting centre of gravity experienced while carrying unevenly distributed loads. The shape of the fin and rudder were also redesigned, with a lead balance weight being fitted to reduce the control forces necessary to move the elevator. In February 1931, the Ju 52 passed the certificate of airworthiness tests at DVL (*Deutsche Versuchsanstalt für Luftfahrt e.V.*)

One very important aspect of the Ju 52 was its *Hilfsflügel* (double-wing). The development of such a device had been the subject of a Junkers 1921 patent (No. 396 621). The *Hilfsflügel* was intended as a twin-aileron assembly, which was to provide a pair of slots as well as flaps when lowered. When the *Hilfsflügel* was lowered prior to landing, the ailerons drooped automatically,

The Ju 52 prototype, registered as D-1974, seen at Tempelhof in late 1930 or early 1931. (*Arlanda Civil Aviation Collection*)

becoming, in effect, flaps. The Junkers T 29 aeroplane, first flown in 1925, was specifically designed to air test the *Hilfsflügel* device.

Although designed primarily as a cargo-carrying aeroplane, it was also thought that the Ju 52 would find favour with passenger airlines. The designation Junkers K 45 applied to a military variant of the single-engined Ju 52. Even though the K 45 was offered to a number of countries, most notably China, none were built. However, as will be described later, one Ju 52 ce was tested as a torpedo carrier in 1933.

In order to increase its export potential, the second Ju 52 (W Nr 4002) was fitted with an 800-hp Armstrong Siddeley Leopard, with the aircraft being designated Ju 52bi. Registered as D-2133, the second Ju 52 was flown for the first time on 10 March 1931. In fact, several different engines were tested on the Ju 52, including the BMW IX U (Ju 52 cai), Rolls-Royce Buzzard (Ju 52cao), and Junkers Jumo 4 (Ju 52do)—the change of engines being indicated by different suffixes. Floatplane tests began on 17 July 1931, with D-2133 being fitted with twin floats. Even before the intensive flight testing had been completed, a series of five Ju 52s was laid down at Dessau. In the event, only four of these single-engined Ju 52s were completed, with the seventh being converted on the assembly line to take three engines. Prior to the first flight of the first three-engined Ju 52, aerodynamic drag tests with dummy engine gondolas had been performed. The first three-engined Ju 52, W Nr 4007, was modified with three Pratt & Whitney engines. First flown in April 1932, the three-engined Ju 52 was designated Ju 52/3m with the 3m indicating *drei motoren* (three engines). The single-engined Ju 52 was retroactively designated Ju 52/1m.

Single-engined Ju 52s

W Nr	Registration/s	Remarks
4001	D-1974, D-UZYP	Prototype. Used for a variety of trials, including DB 600 engine tests
4002	D-2133, D-USUS	Used for Jumo 4 engine trials
4003	D-USON	Fitted with floats. Possibly the Ju 52 used by Luftwaffe as target tug
4004	D-2317, SE-ADM, D-UBES	Used for torpedo trials in Norway in 1933.
4005	D-2356	Used by *Reichsverband der Deutschen Luftfahrtindustrie*, Staaken, Berlin. Dbf 05.1933 following engine fire at Zechlinerhütte
4006	CF-ARM	To Canadian Airways. Remained in service until 1943
4007	D-UHYF	Possibly to Luftwaffe?

Professor Hugo Junkers remarked: 'We have succeeded in building an aeroplane with a large cargo space and yet with outstandingly good aerodynamical qualities'. Indeed, the Ju 52/3m had better performance than its single-engined predecessor, so much in fact that it was decided to discontinue production and development of the latter and instead concentrate on the three-engined aeroplane. Although planned as single-engined Ju 52s, the second production series Ju 52s (W Nr 4008–4012) were all completed as three-engined Ju 52/3ms. One simple, but quite interesting, technical solution was that the engines were—apart from being attached to the engine mount by bolts—also attached with cables. This actually proved its worth early on in the career of the Ju 52/3m.

It must be said that the Ju 52/3m's airframe was not aerodynamically streamlined. Attempts to reduce the built-in drag included fitting NACA or Townend engine cowlings and streamlined wheel spats, the latter reducing drag by 41 per cent. The cabin could be furnished for several different operational requirements. In standard passenger-carrying configuration, seventeen passengers could be carried in relative comfort. For longer scheduled flights, the number of seats was usually reduced to fourteen. One VIP-configured Ju 52/3m was delivered to Romania, and was furnished with comfortable leather seats. At the other end of the spectrum, the Uruguayan airline CAUSA ordered two Ju 52/3ms fitted out with benches seating twenty-six passengers for use on a high-density route. It is interesting to note that the majority of the airlines that ordered Ju 52/3ms used them primarily for scheduled passenger services instead of air freight purposes.

A 1934 advertisement emphasising the cargo capabilities of the Ju 52/3m. (*Author's Collection*)

The inherent strength of the Ju 52/3m's airframe was inadvertently put to the test during an air show near Schleissheim on 1 August 1932. An Udet U 12 light biplane (designed by the First World War-ace Ernst Udet) collided head-on with Ju 52/3m D-2201. The U 12 struck the outboard left engine, severely damaging the wing and fuselage. The port engine was torn from its mount, hanging onto the wing by means of steel cables. The port undercarriage was also torn off. With the U 12 breaking apart, *Flugkapitän* Willi Polte and his co-pilot Achterberg managed to put the severely damaged Ju 52/3m down on a meadow. Aboard the Ju 52/3m were several high-ranking officials, including Erhard Milch, then a director of Lufthansa. Repairs to the Ju 52/3m lasted four weeks.[4]

The superiority of the Ju 52/3m was subsequently demonstrated when D-2201 was entered in a competition with competing aircraft built by Dornier and Fokker. The competition involved flying around the Alps, with the Ju 52/3m breezing through to a comfortable victory.

In the event, the Ju 52/3m was an early success on the export market, with Bolivia and Colombia ordering the type in 1932. That year, other Ju 52/3ms were also acquired by Finnish and Swedish airlines—additionally, one single-engined Ju 52 was sold to Canada. The German national airline Deutsche Lufthansa first purchased Ju 52/3ms in May 1932, continuing a long and successful relationship with Junkers aeroplanes.

After the Nazi's rise to power in January 1933, the *Junkers Flugzeugwerke AG* was nationalised. This was partly a result of Hugo Junkers' well-known antipathy towards the Nazis, but also due to financial issues. Along with other countries, Germany was suffering from the worldwide economic depression. Other factors included the considerable amount of research work performed at Dessau, as well as the design and construction of singular aircraft such as the giant Junkers G 38; Junkers was in desperate need of a successful aeroplane that would appeal to airlines around the world. The airframe and engine companies were eventually merged as the Junkers Flugzeug-und Motorenwerk AG, with Heinrich Koppenberg appointed as Director-General. At the time, Junkers was the largest aircraft manufacturer in the world.

In this context, mention must be made of the Junkers Ju 86 that entered service with Lufthansa in 1936. The twin-engined Ju 86 was the first Junkers airliner to dispense with the corrugated aluminium method of construction. Instead, the Ju 86 was of all-metal monocoque construction. In the event, only a dozen Ju 86s entered service with Lufthansa, with modified variants serving with the Luftwaffe in the bomber and high-altitude reconnaissance roles. On 28 August 1937, another Junkers-designed airliner, the Ju 90 took to the air for the first time. Known as *Der Grosse Dessauer* (The Mighty Dessauer), the four-engined Ju 90 could carry more than twice the number of passengers as compared to the Ju 52/3m. Only thirteen Ju 90s were built, with a developed variant, the Ju 290, seeing service with the Luftwaffe.

Professor Hugo Junkers died on 3 February 1935, the day of his seventy-sixth birthday. He had shown an extraordinarily inventive creativity, filing more than 1,000 patents in fields varying from internal combustion engines, thermodynamics, and all-metal aeroplane structures. As related below, the Ju 52/3m would become one of the world's best known airliners, seeing service on all continents. It would also enter service with the Luftwaffe, initially as a bomber, but also as its main transport aeroplane—used on all fronts. Indeed, without the availability of the ubiquitous Ju 52/3m, the military might of Germany would have resulted in the vital logistics of transporting men and materiel by air severely hampered. Apart from the Luftwaffe, many air arms operated Ju 52/3ms. Post-war, the basic Ju 52/3m entered production in France and Spain.

3
Technical Description

According to the 1930 edition of *Jane's All the World's Aircraft*, the single-engined Ju 52 was a '...commercial cabin monoplane, for freight-carrying', which could be operated on wheels or a twin-float undercarriage. Its power plant was a 720–800-hp Junkers L-88 engine, but any similarly powered air- or water-cooled engine could be used (a number of different engines were fitted to the Ju 52s). The cockpit had two side-by-side seats, and could be fitted with either single or dual controls. Behind the cockpit was the radio operator's seat, the toilet, and 'compartment for post-office or other official'. The single-engined Ju 52 was, to all intents and purposes, a decidedly utilitarian aeroplane. Rugged in appearance, it was designed to carry freight from primitive airstrips. However, the Ju 52 was not a commercial success. Nevertheless, the Ju 52 airframe proved capable of sustained modification, making it one of the world's best known transport aircraft of all time.

Junkers Ju 52 Technical Data and Performance Characteristics

Junkers Ju 52 (Landplane)

Wingspan:	29 m (95 feet, 2 inches)
Length:	18.3 m (60 feet)
Height:	6.3 m (20 feet, 8 inches)
Wing area:	116.5 m^2 (1,255 feet^2)
Empty weight:	3,500 kg (7,700 lb)
Weight (loaded):	6,600 kg (14,550 lb)
Maximum speed at ground level:	200 kph (124 mph)
Service ceiling:	3,200 m (10,500 feet)

The basic configuration for the three-engined Ju 52/3m was, in essence, retained from the earlier Ju 52. Nevertheless, the Ju 52/3m underwent a progressive development throughout its production.

Junkers Ju 52/3m g7e Technical Data and Performance Characteristics

Engine:	Three × 730-hp BMW 132T-2 radial engines
Wingspan:	29.25 m (95 feet, 11½ inches)
Length:	18.90 m (62 feet)

Above: The fuselage of Ju 52 W Nr 4006 taking shape at Dessau. This particular Ju 52 was later sold to Canada and registered as CF-ARM. (*Swedish Aviation Historical Society*)

Left: Wing sections under construction. (*Swedish Aviation Historical Society*)

Note the Junkers-patented wing couplings, which greatly eased assembly and disassembly. Also of interest is the *Werk Nummer* 4006 displayed prominently on a section of the wing. (*Swedish Aviation Historical Society*)

Height:	5.55 m (18 feet, 2½ inches)
Wing area:	110.50 m^2 (1,189.41 feet^2)
Empty weight:	6,500 kg (14,328 lb)
Weight (loaded):	10,500 kg (23,146 lb)
Maximum speed at ground level:	272 kph (169 mph)
Service ceiling:	5,900 m (19,360 feet)
Fuel load:	2,480 litres (545.5 Imperial gallons)
Range:	1,100 km (683 miles)

General Technical Description

The Junkers Ju 52/3m is a three-engined low-wing cantilever monoplane with a wholly enclosed cockpit and passenger/cargo cabin. It is constructed entirely out of duraluminium. The centre section of the wing is integral with the fuselage, with the wings being joined by Junkers patented screw couplings. The wing is built up from eight duraluminium tube spars. The corrugated outer skin stiffens the wing against torsion. A double-wing effect is created through the flap and ailerons hinged on the rear-trailing edge of the wing. They can be operated to increase the wing area, thus increasing the lift of the wing. The ailerons are horn balanced.

The fuselage is built up from duraluminium frames, which together with four longerons form a rectangular fuselage, with domed upper decking. The outer corrugated skin consists of light-alloy sheeting. The tail section is of the normal monoplane configuration, being mounted on top of the rear fuselage and braced by two struts. A fixed-tail wheel is fitted to the lower section of the rear fuselage.

The fixed undercarriage is divided into two separate parts, consisting of half-axles and radius rods hinged to the fuselage. Oleo-pneumatic shock absorbers are attached to the upper wing-root of the forward main spar. Compressed air brakes are fitted. The wheel undercarriage can be replaced by either skis or floats.

The three engines are fitted to the front fuselage (one) and the outer wings (one on each wing). The enclosed cockpit seats two, pilot and co-pilot sitting side-by-side, and dual controls are fitted. The cabin seats between sixteen and eighteen fully equipped troops, or twelve stretcher patients.

4

Ju 52/3m Airliner Operations in Germany

The national airline *Deutsche Luft Hansa AG* (DLH) was an obvious customer for the Ju 52/3m.[1] Formed in 1926 through a merger of Junkers Luftverkehr AG and Deutsche Aero Lloyd AG, the new airline inherited a fleet of 162 aeroplanes, including eighty Junkers-designed airliners. In the event, Lufthansa would become the largest civilian operator of the Ju 52/3m, with the type being introduced into service in May 1932. In all, more than 200 Ju 52/3ms were operated by Lufthansa. Until 1939, the Ju 52/3m established an enviable reputation. During the summer, it had a serviceability record of 97 per cent, which dropped to 90 per cent during the harsh German winters. The accident rate dropped significantly from an average of seven fatal accidents to 1.5 fatal accidents during every millionth kilometre flown. From 1938, this dropped even further to an average of one fatal accident for every millionth kilometre flown.

When, in early 1932, a pair of Rohrbach Roland airliners was transferred to Deruluft, Lufthansa needed replacement aircraft. With the Ju 52/3m trimotor having proved itself, two were ordered from Junkers. The first Ju 52/3m to enter service with Lufthansa was D-2201 (W Nr 4013). First flown on 7 March 1932 by Alfred Gothe, the Ju 52/3m was named *Oswald Boelcke* after the First World War fighter ace. The second Lufthansa Ju 52/3m was registered as D-2202 and named *Richthofen*.

The Ju 52/3m quickly earned a reputation of reliability and ruggedness. Being able to carry a maximum of sixteen passengers comfortably in an enclosed, heated cabin or 3 tons of cargo, the Ju 52/3m was to become the true Lufthansa workhorse of the interwar years. The exact number of Ju 52/3ms operated by Lufthansa is difficult to ascertain, but was probably in excess of 200. Aircraft were transferred to various Lufthansa subsidiaries in China and South America, while others were leased from the *Reichsluftfahrtministerium* (German Air Ministry, RLM) (for a list of Lufthansa Ju 52/3ms, please refer to Appendix I).

The Ju 52/3m was soon operating on the high-profile Berlin to London route, and it was not long before the Ju 52/3m became the standard airliner on Lufthansa's European routes.

In 1933, Lufthansa carried about 95,000 passengers, rising to 130,000 the following year. The quantity of mail carried increased even more, rising by nearly 90 per cent in 1934. This was in no small part due to the increasing numbers of Ju 52/3ms. The introduction of the Ju 52/3m also meant an increase of the average speed of Lufthansa aeroplanes from 160 kph to 240 kph. By 1935, it was expected that the Ju 52/3m would be used on 80 per cent of all Lufthansa routes. This included the joint Russian-German airline Deruluft (described in the USSR section in Chapter 17).

With the arrival of the more capable Ju 52/3m, many of the older, less capacious, and uneconomical airliners were withdrawn from use. One consequence of this was the short regional routes,

The first Lufthansa Ju 52/3m, D-ADOM, seen at Munich airport. (*Hennies Werner*)

A close-up of a Lufthansa Ju 52/3m shortly before take-off from Essen-Mülheim Airport. (*Rob Mulder*)

Hüpsstrecken (literally Jump Routes, the modern equivalent being feeder airline routes) were discontinued. As a result, the number of Lufthansa routes was reduced from sixty to forty-six. The Ju 52/3m had better range, and could also carry more passengers and cargo than the older airliners.

With the European capitals and large industrial centres being connected through a growing network of routes, the Ju 52/3m soon became a common sight at airports around the continent. When compared with the contemporary Douglas DC-2 (and even more with the later DC-3), the Ju 52/3m appeared as a lumbering anachronism, lacking the speed and retractable undercarriage of the sleek Douglas design. With its corrugated steel structure, the Ju 52/3m represented the final development of the 1919-vintage Junkers F.13. However, the Ju 52/3m was rugged, dependable, and able to operate from primitive airfields with little in the way of infrastructure, which could not accommodate the more modern and higher-performance Douglas transports.

Interestingly, Lufthansa issued 'Ten Commandments' for aircraft passengers, which included:

> Ensure that the air in the cabin remains fresh. It is permitted to open the window. It is forbidden to throw any item out of the window as it could fall on the head of the small Earth dwellers who live a long way below you.
>
> After a flight, do not play the role of a hero to your fellow men. For a long time now it has not required special courage to fly. You can render a service to aviation when you enlighten uniformed people who still believe that a flight is a dangerous undertaking.[2]

Some interesting facts can be found within these 'Commandments'. For instance, passengers on domestic flights were allowed 10 kg of luggage, with passengers on international routes being able to bring 15 kg of luggage with them. Smoking was frowned upon, both in the vicinity of the aeroplane when on the ground and in the cabin, 'out of consideration to your fellow passengers'. In order to reduce the engine noise, each passenger was provided with cotton wool for use as earplugs. If a passenger was struck with air-sickness, the advice was to either look out the window, read, or write. Air-sickness pills could, according to the Commandments, be obtained at every major airport.

In 1934, the Transatlantic Airmail Service saw the Heinkel He 70 carrying mail from Berlin to Seville, Spain. Here, the mail was loaded aboard a Ju 52/3m for the onward flight to Bathurst (present-day Banjul) in Gambia. From this African staging point, Dornier Wal flying boats flew to Natal, Brazil. This meant that the Berlin-Rio de Janeiro air mail arrived in a mere four days.

On 28 September 1934, Lufthansa's millionth passenger was carried aboard a Ju 52/3m. In 1935, no less than fifty-one Ju 52/3ms served with Lufthansa; while the following year, Lufthansa had 144 aircraft, fifty-nine of which were Ju 52/3ms. By estimate, the Ju 52/3ms were at the time responsible for 85 per cent of Lufthansa's regular air traffic. While other aeroplanes, such as the giant Junkers G 38 were exceptional in appearance, it was the Ju 52/3m that was the true workhorse of Lufthansa.

Before the inauguration of the Berlin Olympic Games, a Lufthansa Ju 52/3m, D-ALYL, carried the Olympic Torch from Greece to Berlin. For this special flight, the Ju 52/3m was renamed *XI Olympiade*. The Ju 52/3m was later sold to the Austrian airline ÖLAG. Inevitably, a number of Lufthansa Ju 52/3ms were lost in accidents—in 1936, six Ju 52/3ms were written-off. The first of these was D-ASOR, which crashed at Elfingen, Switzerland, on 17 April 1936, killing six crew and passengers—there were only two survivors. Between 19 September and 4 December, Lufthansa lost

A close-up of Ju 52/3m D-AJYR, which was powered by Jumo 205 diesel engines. (*Rob Mulder*)

A Lufthansa Ju 52/3m awaiting the 'clear for take-off' signal at Berlin Tempelhof Airport. (*Rob Mulder*)

four Ju 52/3ms, including D-ASUI, which crashed near Lauf an der Pegnitz while on approach to land at Nuremberg in poor weather.[3]

On longer flights (more than 400 kilometres) with no intermediate stops, the number of passengers was reduced to fourteen. According to an instruction dated 30 April 1937, this was an attempt to improve the level of comfort for the passengers. The ability to travel comfortably by air was also a sales argument.

By 1937, Lufthansa's fleet of Ju 52/3ms flew an average of 1,244 hours for each aircraft, some reaching as much as 1,800 hours, roughly six hours daily. When the famous American aviator Charles Lindbergh visited Germany in 1938, he was quickly put in the left-hand seat of a Ju 52/3m to try out the airliner. Only one Lufthansa Ju 52/3m was lost the year before, D-AGAV. While taking off from Croydon on 26 November 1937 in very poor weather, the Ju 52/3m struck a hangar before crashing.

In the summer of 1937, an attempt was made to open a direct route from Berlin to China. A Ju 52/3m, D-ANOY (W Nr 5663), flown by Carl August Freiherr von Gablenz, Untucht, and Kirchoff, left Berlin-Tempelhof on 14 August, arriving safely in China. In his memoirs, Freiherr von Gablenz wrote about flying the Ju 52/3m at an altitude of 5,400 metres between the Pamir and Karakorum mountains: 'I had the feeling that the wings of D-ANOY were scraping the edges of the cliffs'.[4]

However, on the return flight, the crew was forced down at Kothan. Here, an armed insurrection made it impossible for the crew to proceed, with both crew and aircraft being impounded by the locals. With no news of the overdue Ju 52/3m forthcoming, two Lufthansa aircraft left Berlin, arriving at Kabul, Afghanistan, on 11 September. The flight from Berlin had lasted thirty-one hours, with intermediate stops at Athens and Baghdad to refuel. One of the participants of the search and rescue expedition was Wulf-Diether Graf zu Castell-Rüdenhausen. In all, nine flights were made to try to locate the missing Ju 52/3m. Castell-Rüdenhausen published an account of the rescue flights:

> We turned into the Anjuman Valley and flew northeast at a height of 4,000 metres (13,100 feet) between two mountain ranges rising on either side to 6,000 metres. We were constantly expecting to reach the end of the valley as silhouettes of the mountain ranges came into view, running at right-angles to the valley for it was still too dark to follow clearly the course of the river which wound continually in and out of the mountains. Great care had to be taken not to be misled by these mountains and fly into the valleys; such a mistake might easily have proved disastrous for they were so narrow that turning back would have been impossible. Daylight increased rapidly—day breaks more quickly there than in Europe—and suddenly gigantic pyramids of snow, the 7,000 metre (23,000 feet) peaks of the Hindu Kush, lay around us glistening in the sunlight. The impression was overwhelming. No written description or photograph could convey the grandeur of the scene.
>
> At the end of an hour we crossed the Anjuman Pass at an altitude of nearly 5,000 metres (16,400 feet). In the meantime the clouds had increased considerably; our valley was still clear, but they clung to the mountains to the right and left of us. The pass leading to the Wakhan Valley was already in cloud, but according to my map there was another route and this I determined to find. I reached the other valley through a maze of rocky peaks. Now and again the snow-clad summits of an 8,000 metre (26,250 feet) giant could be seen rising above the clouds and I attempted to impress on my mind the shapes of the most prominent peaks so that they would serve as landmarks for the return journey. The increasing cloud no longer allowed a clear view and made flying difficult. The aircraft would not climb above 5,000 metres with the load aboard and we were forced to fly along the lower edge of broken cloud.

> For two hours more we continued to fly between the mightiest glaciers in the world, endless snowfields, and boulder-covered slopes; then suddenly it all came to an end. Immediately in front of the Wakhan Pass was a cloud bank and we had to turn back. It was quite time for us to begin the return journey, for the higher the sun rises and the warmer the air becomes, the lower the clouds descend into the valleys. As we approached the 5,000 metres pass which we had to cross to reach the Anjuman Valley, there was only a small opening between the clouds through which we could fly. With such bad visibility great caution was necessary to avoid a fatal turn; we landed at Kabul about midday.[5]

On 24 September, the missing Ju 52/3m arrived unannounced over Kabul airport, causing great joy among the German rescuers. The three Ju 52/3ms arrived back at Tempelhof to a tumultuous reception. In the event, the scheduled Far East service was inaugurated on 29 October 1937. Initially, the route ended in Baghdad. A few months later, it was extended to Tehran and Kabul. Two brief visits were made to Japan in May and August 1939; however, the outbreak of war hindered further development of the Lufthansa Far East route.

5

Lufthansa Around the World

Lufthansa influence meant that a number of affiliate companies in different parts of the world, particularly in South America, were formed. With Lufthansa establishing control over the airways and serving as an important propaganda tool, this was viewed with an increasing level of mistrust by the US government. In fact, following the outbreak of war on 1 September 1939, the US would apply pressure on the South American governments where German-controlled airlines were operating to reduce German controlling interests. Interestingly, the FBI would charge SEDTA with transporting secret agents and arms to rebel groups in the region, using the airline's Ju 52/3ms. This remains unsubstantiated, though.[1]

Most of these airlines, including LAB in Bolivia, Syndicato Condor, VARIG and VASP in Brazil, SEDTA in Ecuador, and Sucursal Peru in Peru, operated Ju 52/3ms. Lufthansa also formed the *Lufthansa Bezirksdirektion Südamerika* (South American Regional Management), with five Ju 52/3ms being delivered, either from new production or from Syndicato Condor. Registered as D-ARUW, D-AGST, D-AENF, D-APEF, and D-AMYE, all were eventually transferred to Syndicato Condor. The first Lufthansa flight between Buenos Aires and Rio de Janeiro was made on 13 September 1937, using Ju 52/3m D-AENF.

That particular aircraft had been transferred from Syndicato Condor on 31 August 1937, and re-registered as D-AENF. The aircraft was painted in Lufthansa colours, although the Brazilian name Aconcagua was retained. The Ju 52/3m made frequent visits to Ecuador and Peru as well as Argentina. Following an earthquake in Chile, the Ju 52/3m took part in the supply of emergency aid. This particular aeroplane would go on to become one of the most famous of all Ju 52/3ms, probably holding a record regarding the number of registered owners.

In this context, it should be mentioned that not all German civilian-registered Ju 52/3ms were operated by Lufthansa. Many aircraft served with the *Deutsche Verkehrsfliegerschule GmbH* in Berlin, operating Ju 52/3ms in training airline pilots. Several companies also used Ju 52/3ms, including Hansa Luftbild (D-AGDE). Other civilian registered Ju 52/3ms were used for a variety of tests by Junkers and the Erprobungsstelle Rechlin, including engines and different kinds of equipment. On the eve of war, eighty Ju 52/3ms were on charge with Lufthansa, out of a fleet of 145 aircraft.

Until August 1939, Lufthansa operated the London–Hamburg–Copenhagen–Gothenburg–Oslo route (Route 27) using Ju 52/3ms.

The number of Ju 52/3m serving with DLH fluctuated, particularly following the outbreak of war in 1939. Many aircraft were transferred to the Luftwaffe, while others were leased or sold to foreign airlines such as Aero O/Y and Iberia. The Berlin–Copenhagen–Malmö route (Route 5) was flown daily during the war in cooperation with ABA and DDL, with the Berlin–Stockholm route (Route 8) was flown six days per week (not on Sundays) in cooperation with ABA.

In 1940, seventy-eight Ju 52/3ms were on charge with Lufthansa. However, several of these were used by the *Fliegerstaffel des Führers*, with others being operated for experimental purposes.

By 1942, most of Lufthansa's aircraft had been pressed into Luftwaffe service, with others being lost to enemy action and accidents. Only fifty-nine aircraft remained, forty-seven of which were Ju 52/3ms. The final Ju 52/3m to be delivered new to Lufthansa was a Bernburg-built aircraft, D-AIAH (W Nr 7268). Taken on charge on 29 January 1942, it was transferred to Luftwaffe less than three weeks later, on 18 February 1942.

One important task for the Lufthansa Ju 52/3ms was the delivery of mail. Receiving a letter from your wife or girlfriend could work wonders with your fighting morale. With German forces on the retreat on all fronts, the number of routes operated by Lufthansa diminished accordingly. Sweden and Spain were still among the destinations, as was a route along the Norwegian coast. On 31 December 1944, thirty-one Ju 52/3ms remained with Lufthansa. However, out of these, twenty-eight were leased to the Luftwaffe. The final Lufthansa Ju 52/3m to be lost during the war was D-ANAJ. Following take-off from Berlin Tempelhof on the final Lufthansa scheduled flight out of the beleaguered German capital, the Ju 52/3m was shot down by Soviet forces at Gilenig. The intention had been to make an intermediate stop at Munich before continuing to Prague. Everyone aboard, including the famed movie director Hans Steinhoff, was killed.

On 23 April, the final Lufthansa flight departed Berlin Tempelhof. The airport was then being continuously shelled by Soviet artillery. Subsequent Lufthansa flights to Sweden were made from Warnemünde on the Baltic Sea coast. The final Lufthansa flight to Sweden occurred on 3 May 1945, when D-AUAV landed at Bulltofta. No passengers were aboard, however, some mail bags were carried. In the event, D-AUAV returned to Germany, being surrendered at Flensburg. Following Hitler's suicide on 30 April, the German government relocated to Flensburg. On 3 May 1945, Ju 52/3m D-AFFF arrived back at Flensburg having flown a courier flight between Aalborg and Oslo. Canadian troops entered Flensburg on 6 May, with the war finally coming to an end on 8 May. In all, some 200 Lufthansa aircrew lost their lives during the Second World War.

6
Ju 52/3m Production in Germany

The Ju 52/3m entered limited series production in 1932, and the Junkers production line at Dessau built aircraft for both indigenous customers and for export. Demand for the Ju 52/3m soon outgrew output at Dessau, when the early 1930s saw a large expansion of the German aero industry. On 13 December 1933, the so-called ABC Programme resulted in an order for 179 Ju 52/3ms being placed, with the aircraft mainly intended for the still-secret Luftwaffe. This large order was divided among four companies: Junkers, ATG at Leipzig, Dessauer Waggon at Dessau, and Hamburger Flugzeugbau at Bremen, with the first two handling final assembly. Dessauer Waggon and Hamburger Flugzeugbau were to be responsible for component manufacture. One aim for the ABC Programme was to introduce the methods of all-metal aircraft construction to other German aircraft manufacturers than Junkers.

ATG

Located at Leipzig in eastern Germany, the *Allgemeine Transportanlagen-Gesellschaft* (ATG) had originally been established in 1911 as the *Deutschen Flugzeug-Werke* (DFW). Following the end of the First World War, the company was reformed as *Allgemeine Transportanlagen-Gesellschaft GmbH*. ATG was controlled by Heinrich Koppenberg, which considerably eased negotiations of Ju 52/3m production. Final assembly took place at Leipzig-Mockau, with component production being located to Leipzig-Lindenthal. Tooling up lasted about one year, with 32 Ju 52/3ms being assembled by ATG in 1935; the following year, seventy-two Ju 52/3ms were built. Between 1935 and 1937, ATG delivered 154 Ju 52/3ms, but production was discontinued in 1937, eventually being restarted in 1940 due to an increased demand for transport aircraft, lasting until mid-1944.

ATG Ju 52/3m Production

Year	Quantity built
1935	32
1936	72
1937	50
1940	120

1941	201
1942	204
1943	258
1944	101
Total:	1,038

The *Werk Nummer* batches allocated to ATG-built Ju 52/3ms were 2801 to 3462, and 500000 to 500xxx and 501000 to 501xxx.

Junkers

The main producer of the Ju 52/3m was, of course, Junkers. Between 1932 and 1944, Junkers built nearly 3,200 Ju 52/3ms. Initially, many aircraft were exported, earning much needed revenue for the cash-starved company. One of the most vocal opponents to Erhard Milch's organisational visions for the German aeronautical industry was Hugo Junkers. This led to a bitter conflict between Junkers and Milch, with the latter putting forward accusations of economic mismanagement, particularly with regards to the factory at Fili. In late 1931, Junkers received a harsh directive to put the company's finances in order. In December 1931, Klaus Junkers, Hugo Junkers' son, took charge of the *Junkers Flugzeugwerke*; in the event, the Junkers companies were nationalised in late 1933. At one point, Hugo Junkers even risked facing trial as a traitor to the German state. On 24 November 1933, Heinrich Koppenberg was appointed Director of Junkers. Inspired by US ideas of industrial mass production, Koppenberg modernised the production line at Dessau, as well as licensing construction of the Ju 52/3m to ATG at Leipzig. A new production programme, named ABC, was introduced. The ABC programme provided master rigs for components manufacture, which were then delivered to sub-contractors across Germany.[1]

Following the nationalisation of Junkers in 1933, the Ju 52/3m was selected as one of the most important types of aircraft for the still-secret Luftwaffe. In July 1935, Junkers employed 9,483 airframe workers and 5,435 workers in the engine department. The company itself was valued at 130 million RM. A second Ju 52/3m production line was set up at Bernburg, with the first Ju 52/3m built at Bernburg rolling off the production line on 2 October 1937.

In all, 3,192 Ju 52/3ms were built by Junkers, with production at Dessau ending in 1938 with the Ju 52/3m g4e variant. Ju 52/3m production at Bernburg lasted between 1937 and 1944. Early in the war, component production was dispersed. With the Dutch aircraft manufacturer having refurbished a large number of Ju 52/3ms damaged during the invasion of the Netherlands, the company received an order for 200 sets of floats (only sixty-six pairs of floats were completed) as well as 685 engine blocks and sixty engine mounts—the latter were not completed.

Located in the Reichsprotektorat Böhmen und Mähren (the present-day Czech Republic), the company Letov was assigned the production of Ju 52/3m wings. The final Junkers-built Ju 52/3ms were delivered to Luftwaffe in the autumn of 1944. In early 1945, the *Junkers Flugzeug- und Motorenwerk AG* controlled no less than eighteen plants, having some 140,000 employees.

Junkers-built Ju 52/3ms

Year	Quantity built
1932	12?
1933	17
1934	175
1935	347
1936	221
1937	210
1938	300
1939	578
1940	303
1941	301
1942	329
1943	318
1944	81
Total:	3,192

The following *Werk Nummer* batches were assigned to the Junkers-built aircraft: W Nr 4001 to 4080, W Nr 5001 to 6000, and W Nr 6001–7800.

Weser Flugzeugbau AG

Although no complete Ju 52/3ms were built by *Weser Flugzeugbau*, which had been formed in April 1934, this company was given a contract to convert obsolete Ju 52/3m g3e and g4e bombers to transport configuration. This work was initiated in 1937, with twenty-one aircraft having been completed by 1 April 1938. Orders were received for the conversion of an additional 383 Ju 52/3m bombers, to be completed by April 1939 (i.e. an average of thirty-two aircraft each month).

Deliveries to Luftwaffe

The exact number of Ju 52/3ms delivered to Luftwaffe is unclear, but it is probable that about 4,400 aircraft were taken on charge between 1933 and 1944. In this context, it is important to remember that the actual quantity produced does not always tally with the quantity accepted by the Luftwaffe. Aircraft destroyed in pre-delivery crashes, aircraft destroyed in bomb raids, transfers to other operators, or aircraft allocated by the manufacturer for various tests must be taken into consideration. Nevertheless, from available sources it would appear that 2,804 Ju 52/3ms were accepted by the Luftwaffe between September 1939 and September 1944. Quantitatively, the Ju 52/3m was by far the most important type of transport aeroplane in Luftwaffe service. In comparison, a mere thirty-one Ju 352s, forty-three Gotha Go 244s, and 201 Messerschmitt Me 323s were delivered to the Luftwaffe *Transportverbände*.[2]

In 1943, no less than 900 Ju 52/3ms were delivered, including Junkers, ATG, and French production. Between January and July 1944, 376 Ju 52/3ms were accepted; however, by 1944,

Germany was being attacked by RAF bombers by night and USAAF bombers by day. In an aircraft priority list, Ju 52/3ms and other transport aircraft were placed at number seven. Fighters required to defend the Reich had the highest priority. Ju 52/3m production ended in September 1944; its intended replacement, the Ju 352, never entered large-scale production.

Luftwaffe Ju 52/3m deliveries, September 1939 to September 1944

Year	Quantity
September–December 1939	145
1940	388
1941	502
1942	503
1943	887
1944	379
Total:	2,804

The exact number of Ju 52/3ms built is difficult to establish. Between September 1939 and 1944, a total of 3,225 Ju 52/3ms were built. Taking production by all manufacturers into account, the number of Ju 52/3ms built is 4,768, divided as follows: Junkers (3,192), ATG (1,038), Amiot (538), and PIRT (around thirty). Post-war production in France and Spain added 415 and 170 aircraft respectively. Thus, total production could be as high as 5,353 aircraft.

7
Foreign Production

France

Following the occupation of France, the production of several types of German aircraft were transferred to French factories. In late 1940, Amiot was assigned the production of Ju 52/3ms at its Colombes factory. The production of the various sub-assemblies was dispersed, with Amiot being responsible for the fuselage and wing centre section, SNCASO at Suresnes for the outer wings with its factory at Courbevoie producing the tail section and undercarriage. Assembly and flight testing was undertaken by Breguet at Vélizy-Villacoublay. Tooling up for production lasted more than eighteen months, with the first French-built Ju 52/3m being delivered to the Luftwaffe in June 1942. By the year's end, forty Ju 52/3ms had been built in France. In 1943, 321 were built, and 538 Ju 52/3ms were built in total by Amiot in 1942–1944.[1]

Following liberation, Amiot was nationalised as Ateliers Aéronautiques de Colombes (AAC). The Ju 52/3m production line remained open, with no less than 415 Ju 52/3ms built post-war under the local designation AAC.1. Most of the early production aircraft went to Air France, with the first aircraft being handed over in December 1944. Apart from Air France, a small number of charter companies used the AAC.1. However, the majority of the AAC.1's were delivered to the *Armée de l'Air*, with others being taken on charge by the *Aéronautique navale*. The final AAC.1 was delivered in 1948. In French service, it was known as the Toucan (Tucan). Although the AAC.1 had largely disappeared from the airways by the early 1950s, the final *Armée de l'Air* AAC.1 was struck off charge in March 1962, with the *Aéronautique navale* retaining their Toucan's until 1963. A few AAC.1s were exported to Lebanon, Portugal, Syria, and Yugoslavia.

Hungary

According to the German-Hungarian Aircraft Production Agreement, a number of German-designed aircraft would be built in Hungary. The completed aircraft were to be divided between the Luftwaffe and RHAF. Negotiations for the production of Ju 52/3ms in Hungary resulted in the Budapest-based *Pestszentlőrinci Ipartelepek RT* (PIRT) building about thirty aircraft. Established on 7 October 1942, PIRT received its first order from the Hungarian government three weeks later, on 29 October, for 100 Ju 52/3ms. In March 1943, the RLM placed an order for 160 Ju 52/3ms. According to the planned delivery schedule, thirty Ju 52/3ms were to be completed each month. In July 1943, the RLM ordered an additional eighty-eight Ju 52/3ms. In the end, the harsh realities of war meant that only about thirty Ju 52/3ms were actually built by PIRT.

The first three PIRT-built Ju 52/3ms had been completed by the end of 1943 with another twenty-seven being delivered the following year. Other sources state that thirty-four Ju 52/3m g8es and g14es were built in 1944, with another two airframes uncompleted. The armament consisted of a flexible 7.92-mm MG 17 or 12.7-mm GMK machine gun mounted in the open dorsal position. Additionally, a pair of 7.92-mm MG 17 machine guns was fitted in the sides of the fuselage. Due to the pressing need for combat aircraft, production of the Ju 52/3m ceased in 1944. Four of the Ju 52/3ms were supplied to Luftwaffe, with the remainder being delivered to the Royal Hungarian Air Force (RHAF). Most of the Ju 52/3ms were destroyed during the final months of the war.

Spain

The Spanish company *Construcciones Aeronáuticas Sociedad Anónima* (better known under the acronym CASA) initiated licence production of the Ju 52/3m in 1944. However, already in 1937, *Teniente Coronel* Felipe Lafita had suggested that the Ju 52/3m should be built in Spain as a standard transport aeroplane. However, it was not until after the nationalist victory that these plans were realised. On 28 November 1940, negotiations were initiated to obtain the rights to produce the Ju 52/3m g7e under licence. A contract for the delivery of 100 aircraft was signed on 23 October 1941. It was intended that the first five would be delivered in 1942, followed by forty in 1943 and the final fifty-five in 1944. In the event, the first CASA 352, as the licence-built model had been designated, did not roll off the production line until mid-1944, being taken on charge by the *Ejército del Aire* (EdA) on 6 July. Production was centred at CASA's facilities at Getafe, 10 km (6 miles) south of Madrid.

The CASA 352 was based on the Ju 52/3m g10e. The radio equipment chosen was the FuG X. Regarding a choice of engines, it was found that 149 BMW 132As, twenty BMW 132Ns, and no less than 269 M-25 engines were available in Spain. The basic M-25 was the Pratt & Whitney R-1820

A view of the CASA factory at Getafe, with a number of CASA 352s awaiting delivery. (*Luis Gonzalez Pavón*)

built under licence in the USSR and Spain. Although many M-25s were available, those were needed for the Polikarpov fighters still in service with EdA. One further complication was the lack of spares. In the end, ENMASA B-3 engines were fitted to the first batch of CASA 352s.

An amended schedule called for one to be completed in 1944, ten in 1945, twenty in 1946, thirty in 1947, and forty in 1948—this was not achieved either. In the end, the final aircraft of the first contract was not delivered until 1951. The EdA placed orders for a further seventy CASA 352s, with the contracts being signed on 14 October 1948 (thirty aircraft) and on 11 March 1952 (forty aircraft). The final four aircraft were not to be delivered until 1955, with the last one, s/n T.2B-279, being taken on charge on 10 June 1955.

The CASA 352 was produced in three main variants: A-1, A-3, and C. The CASA 352A-1 was a paratroop variant (125 built), the A-3 was a passenger transport (fourteen built), and C (thirty-one built) was a dedicated training platform for navigators and radio operators. A few aircraft were converted in service, mostly A-1 to A-3 or *vice versa*.

The B-3 engines were eventually replaced by the ENMA Beta B-4 engine. Built by Empresa Nacional de Motores de Aviación S.A. (ENMA) the nine-cylinder air-cooled supercharged Beta B-4 was rated at 775 hp at 2,200 rpm for take-off, with cruise rating being 550 hp at 1,600 rpm at an altitude of 3,000 metres. The last CASA 352s were withdrawn from use in the late 1970s, and many have survived into preservation.

Variants

Although the Ju 52/3m was continually developed during its twelve-year production run, the basic airframe remained essentially similar. The first two single-engined Ju 52s, *Werk Nummer* 4001 and 4002, were powered by a succession of engines: Junkers L-88 (designated Ju 52 ba), BMW VIIaU (Ju 52 be), and Armstrong Siddeley Leopard (Ju 52 bi). The first letter, 'b', indicated the first production series with the second letter indicating the type of engine. These aircraft had a maximum take-off weight of 7,000 kg.

The following five Ju 52s, *Werk Nummer* 4003 to 4007, featured a slightly modified airframe, as well as an increased maximum take-off weight (7,200 kg). These Ju 52s had the suffix letter 'c', indicating the second batch, with a second and, in two cases, a third letter indicating the type of engine. Apart from the Junkers L-88 (Ju 52 ca), BMW VIIaU (Ju 52 ce), and AS Leopard (Ju 52 ci), the BMW IX U and Rolls-Royce Buzzard engines were also fitted, with the aircraft being designated Ju 52 cai and cao respectively.

This system was continued with the first Ju 52/3m series, with da, de, di, and do sub-variants.

Main Ju 52/3m Variants

Type	Engine	MTOW	Remarks
Ju 52/3m de			W Nr 4008 to 4012. First batch of Ju 52/3ms. Exported to Bolivia and Colombia.
Ju 52/3m ce	3 × 525-hp BMW Hornet A	9,200 kg	Delivered to Lufthansa and for export.
Ju 52/3m ba	3 × 575/750-hp Hispano-Suiza 12		W Nr 4016 CV-FAI for Prince Bibescu.

Ju 52/3m fe	3 × BMW Hornet A-2	9,200 kg	Improved model featuring wheel spats and NACA cowlings on outer engines. Fifteen built.
Ju 52/3m ge	3 × 525-hp NMW Hornet for early aircraft 3 × 660-hp BMW 132 A	10,000 kg	Export variants included 'sai' and 'vai', both intended for Swedish Airlines.
Ju 52/3m ho	3 × Jumo 205 diesel engines		Trials only. Two, D-AJYR and D-AQAR briefly used by Lufthansa.
Ju 52/3m Z	3 × 880 h.p. BMW 132 Dc	12,500 kg	Improved cabin comfort. Three-bladed constant speed propellers. A few delivered to DLH and export customers.
Ju 52/3m g3e	3 × 725-hp BMW 132 A-3		Tailskid replaced by tailwheel on late production aircraft. Could carry 1,500-kg bombs, armed with two MG 15 machine guns. Used in combat in Spain.
Ju 52/3m g4e	3 × 525-hp BMW 132 A		Dedicated transport variant, similar to 'g3e'. Had tailwheel instead of tailskid. No lower gun turret, armed with one dorsal-mounted MG 15 machine gun. Fitted with float attachment points.
Ju 52/3m g5e	3 × 830-hp BMW 132 T-2	10,500 kg	Similar to 'g4e'. Fitted with float and ski attachment points
Ju 52/3m g6e	3 × 830-hp BMW 132 T-2		Specialised model for transport duties with extra radio gear and automatic pilot. Could be fitted with degaussing ring. Provision for two 7.92-mm machine guns.
Ju 52/3m g7e	3 × 830-hp BMW 132 T	11,000 kg	Transport model for eighteen troops or twelve stretchers. No wheel spats. Larger cargo doors. Armed with one dorsal-mounted MG 131 and two MG 15 firing through open side windows. An autopilot was also fitted.
Ju 52/3m g8e	3 × 830-hp BMW 132 T, later variants had 3 × 850-hp BMW 132 Z		Similar to 'g6e' with increased radio fit and direction finding gear. A few fitted with 11,000-litre floats.
Ju 52/3m g9e Ju 52/3m g4e (Trop)	3 × 850-hp BMW 132 Z	11,500 kg	Tropical variant intended for service in North Africa. Reinforced armament. Fitted with glider towing gear and strengthened undercarriage.
Ju 52/3m g10e	3 × 850-hp BMW 132 Z	10,500 kg (land) 11,000 kg (sea)	Land or sea transport model. Similar to 'g9e', lacked de-icing gear. Armament as 'g7e'.

Ju 52/3m g12e	3 × 830-hp BMW 132 L		Land transport model
Ju 52/3m 12	3 × 830-hp BMW 132 L		Civilian model of Ju 52/3m g12e for DLH.
Ju 52/3m g14e	3 × 830-hp BMW 132 L		Final production variant. Increased armour protection. One MG 15 mounted in forward dorsal streamlined cupola.

A number of specialised applications were available through seven different *Standard-Bausätze* (literally standard conversion kits), identified through a suffix letter (e.g. Ju 52/3m g8e(F)); the kits are listed below:

(F) *Fallschirmjäger und Luftlande Transporter* (Parachute & air supply transport)
(H) *Hörsaal Flugzeug* (Flying lecture hall, used to train navigators and radio operators)
(K) *Kistentransport* (Dedicated cargo transport)
(N) *Nachtschub Transport* (Dedicated night cargo transport)
(R) *Reise Flug* (Courier aeroplane)
(S) *Sanitätsflugzeug* (Ambulance aeroplane)
(St) *Staffeltrupp-transport* (Squadron Level Transport)

Following the General Francisco Franco's rebellion against the Republican government on 18 July 1936, Germany and Italy quickly offered support to the nationalist rebels. Hitler, with the complete support of the Luftwaffe's Commander-in-Chief Hermann Göring, promised that a volunteer unit would be sent to Spain, along with deliveries of aircraft and other military equipment. The Spanish Civil War would eventually become a proving ground for new German and Italian combat aeroplanes, as well as for developing air combat tactics. Among the first German aeroplanes to arrive in Spain were a few Ju 52/3ms, being used both as transports as well as bombers. The first German Ju 52/3m to actively assist the rebel cause was a Lufthansa example. Returning from West Africa in D-APOK (W Nr 5074), *Flugkapitän* Alfred Henke and his crew made an intermediary landing in the Canary Islands on 20 July. Here, Franco forces requested that the Germans fly General Luis Orgaz to Tétouan in Spanish Morocco. The German crew duly complied, arriving at Tétouan on 22–23 July. At Tétouan, an improvised airlift operation to Metropolitan Spain was organised.

Henke subsequently arrived back in Berlin on 24 July, carrying a Nationalist emissary. Hitler initially refused to send troops, stating that only aircraft would be involved. On 26 July, it was decided to despatch a number of Ju 52/3ms to Spain. The highly secret operation was named Operation *Feuerzauber* (Operation Magic Fire). Henke was assigned to fly the first Ju 52/3m, which had been stripped of its passenger seats. Thirty-five soldiers were able to sit, albeit very tightly packed, on the floor of the aeroplane. Two Lufthansa Ju 52/3ms were used during the latter stages of the airlift, transporting 2,489 troops in two weeks. To reduce the risk of being hit by republican naval vessels, the flights were usually made at altitudes varying between 2,500 and 3,500 metres. The airlift of troops from Spanish Morocco ended on 11 October, with 13,900 troops and 270 tons of supplies being carried in 868 sorties. Only one Ju 52/3m was lost, being written-off in a crash at Jerez on 15 August.[2]

During this operation, nine Ju 52/3ms were initially sent to Spain, with the aircraft being drawn from I/KG 153 at Merseburg, II/KG 155 at Neukirchen-bei-Ansbach, and I/KG 253 at Gotha. An additional

Forward-firing machine guns were fitted to Ju 52/3m s/n 22-90. (*Luis Gonzalez Pavón*)

eleven Ju 52/3ms were eventually ferried to Morocco. The Luftwaffe aircraft had been painted in Lufthansa colours, with one landing at Badajoz through navigational error, falling into the republicans' hands. The crew was eventually released, with the Ju 52/3m being destroyed in an air raid.

Meanwhile, a number of Luftwaffe bomber crews had arrived in Spain. The German Commander, *Oberleutnant* Rudolf Freiherr von Moreau, issued orders that two Ju 52/3ms to be modified as bombers. In 13 August 1936, the republican battleship *Jaime I*, which had been observed in the Bay of Malaga, was seriously damaged by two SC 250 bombs, killing forty-seven sailors. In the event, an additional four Ju 52/3ms were modified as bombers. On 7 November 1936, the German volunteer unit *Legion Condor* arrived in Spain. The *Legion Condor* consisted of air, ground and naval forces, including *Kampfgruppe* 88 (K/88), which was initially equipped with thirty-one Ju 52/3m g3e and the g4e variants. The Ju 52/3ms were divided among three *Staffel*. Early on it was seen that the defensive armament of the Ju 52/3m was inadequate. Republican fighters would usually attempt to bring down a Ju 52/3m by making head-on attacks, and, as a result, machine guns were fitted in the wings, as well as a front turret—a simple aiming device was also mounted in front of the cockpit. On 4 November, republican fighters attacked a *Legion Condor* Ju 52/3m, injuring the gunner *Leutnant* Oskar Kolbitz. Trapped in the lower gun position, the crew were unable to pull Kolbitz into the aircraft. When the Ju 52/3m landed, he was crushed to death.

A number of bomb raids were flown by K/88 in late 1936 and early 1937, and one such raid would achieve notoriety. On 26 April 1937, K/88 set out to raid a railway station and bridge some 20 km east of Bilbao, code-named Operation Rügen. Due to poor weather, the nearby town of Guernica became the target of the bombers instead. About 300 people were killed, with 70 per cent of the town levelled. A mix of high-explosive (HE) and incendiary bombs were used during the raid. The raid against Guernica caused international furore, with the city of Guernica becoming synonymous with terror bombing against civilian targets. One eyewitness account stated that the Basque forces and civilian population had learned to differentiate between 'white' and 'black' bombers, with the latter being Ju 52/3ms:

> The 'white' aeroplane made a reconnaissance over the town, and when directly over the centre of it, dropped three bombs. Immediately afterwards he saw a squadron of seven aeroplanes followed a little later by six more, and this in turn by a third squadron of five more. All of them were Junkers. Meanwhile Guernica was seized with a terrible panic.... The aeroplanes left around 7 o'clock, and then there came another wave of them, this time flying at immense altitude. They were dropping incendiary bombs on our martyred city. The new bombardment lasted 35 min, sufficient to turn the town into an enormous furnace.[3]

Although a number of additional bomb raids were flown, the Ju 52/3m bombers had been phased out by the summer of 1937, and replaced by Dornier Do 17E-1 and Heinkel He 111B-1s. The *Legion Condor* received fifty-nine Ju 52/3mg3e and g4es in total, and two floatplanes were included in this number. By August 1937, most of the remaining *Legion Condor* Ju 52/3ms were transferred to the Nationalist air arm. Fifty-five Ju 52/3ms were transferred to the Nationalist air arm, including non-*Legion Condor* aircraft.

Leutnant Peter Boddem, who had scored ten aerial victories in Spain, was killed in the crash of a Ju 52/3m while returning to Germany in March 1939.

The day before the annexation of Austria (known as Anschluss) on 13 March 1938, considerable numbers of Ju 52/3ms dropped leaflets over towns and villages, informing the Austrian population of their impending incorporation into the German Reich. On 13 March, paratroopers were flown to Graz, while Bf 109s of 1./JG 132 escorted Hitler's Ju 52/3m to Vienna.

During the Anschluss, the subsequent occupation of Memel in Lithuania and the Sudetenland, not a single shot was fired. Nevertheless, for the Ju 52/3m crews, these operations provided operational training for things to come. *Kampfgruppen* b z b V 1 and 2 were soon to be incorporated into the *Kampfgeschwader* b z b V 1.

Poland

When German forces invaded Poland on 1 September 1939, Luftwaffe units fielded 1,580 aeroplanes, including eleven *Transportgruppen* with 536 Ju 52/3ms, 499 of which were serviceable; however, only 120 of these were dedicated cargo transports. Not included in this total were the twelve Ju 52/3ms (ten serviceable) night bombers of 1./Ln. Abt.100. Additional Ju 52/3ms served with various second-line units. Losses were relatively light, with thirteen Ju 52/3ms being written-off (60–100 per cent damage) to all causes with another twenty-two (or thirty-two) suffering 20–59 per cent (repairable) damage. During the September campaign against Poland, the *Transportgruppen* delivered supplies, including fuel, food rations, and munitions, to the advancing forces. Some units of the Wehrmacht's 22. Infanterie-Division were airlifted to Lodz, which proved to be the only proper airlift operation during the Polish campaign.

The first Luftwaffe Ju 52/3m loss (listed as a 'Ju 52/3mT') during the Second World War occurred on 3 September, when an aircraft of 1./St.G.1 (a Ju 87 Stuka unit) crashed on landing at Gryźliny in foggy conditions. Two days later, a civilian-registered aircraft pressed into service, D-ASCG, was shot down in error by the cruiser *Admiral Scheer*.

On 15 September, most of the Ju 52/3m transport units were consolidated into a strengthened Wing, including III./KG z b V 1, IV./KG z b V 1, half of II./KG z b V 2, III./KG z b V 2, IV./KG z b V 2, and KG z b V 172. On the same day, the first Ju 52/3m bomb raid took place when five aircraft dropped a total of 2,370 1-kg fire bombs over Luszyn, where a Polish Army staff headquarters was located.

Ju 52/3ms of IV./KG z b V 2 also took part in the 25 September raid on Warsaw. Having temporarily deployed to Kacin, some 45 km south of the Polish capital, three separate raids were flown against Warsaw on 25 September, with 103,510 1-kg fire bombs being dropped—each aircraft carried 1,200 such bombs. Two aircraft, coded G6+CZ and G6+GZ, were shot down by Polish flak. The following day, IV./KG z b V 2 returned to Breslau-Gandau. (When the Jews of the Warsaw ghetto rose up in April 1943, Luftwaffe Ju 52/3ms were put into action as bombers to quell the rebellion. This was the second and last time that Ju 52/3ms had been used for such tasks during the Second World War.)

Luftwaffe Ju 52/3m Transport Units: 1 September 1939

Unit	Base	Total number of aircraft	Serviceable aircraft
Stab/KG z b V 1	Liegnitz (Legnica)	3	3
I./KG z b V 1	Schönefeld–Seifersdorf	53	53
II./KG z b V 1	Schönefeld–Seifersdorf	53	53
III./KG z b V 1	Aslau	53	52
IV./KG z b V 1	Liegnitz & Lüben	53	53
Stab/KG z b V 2	Küpper-Sagan	3	3
I./KG z b V 2	Sorau (Żary)	52	52
II./KG z b V 2	Freiwaldau (Jeseník)	53	53
III./KG z b V 2	Freiwaldau	52	52
IV./KG z b V 2	Breslau-Gandau	53	53
Stab/KG z b V 172	Berlin Tempelhof Airport	3?	3?
II./KG z b V 172	Berlin Tempelhof Airport	26	16
III./KG z b V 172	Berlin Tempelhof Airport	26	0
KGr z b V 9	Aslau	53	53

The winter of 1939–40 was very cold, which meant that islands along the German North Sea coast became isolated. Floatplanes could not be used, with three Ju 52/3ms, D-TABW, D-TABX, and D-TABY of *Sanitätsflugbereitschaft* 11, flying a total of 214 sorties, carrying 1,284 passengers and 108,629 kg of cargo (including mail).

Denmark and Norway

The Nordic countries had initially been spared from the war. However, when Soviet forces attacked Finland on 30 November 1939, the situation deteriorated rapidly. The small Finnish army and air force fought bravely against the numerically superior Soviet forces. In early 1940, it was proposed that a joint Anglo-French Expeditionary Force would travel to Finland by means of crossing

The Wehrmacht was dependent on the Luftwaffe providing necessary airlift capability. (*Author's Collection*)

Norway and Sweden. Not mentioned to the Norwegian and Swedish governments was the fact that the main operational directive of the Expeditionary Force was to seize the Swedish iron ore mines, along with its railway transport routes. Securing the supply of the high-grade Swedish iron ore was deemed vital to both Britain and Germany.

However, permission for the Expeditionary Force to enter Norway and Sweden was denied. Iron ore from neutral Sweden continued to flow into the German war industry, although only about one-third of it was being shipped through the Norwegian city of Narvik. Nevertheless, a directive issued on 1 March concerned the preparations for a German invasion of Norway. The invasion was to be known as Operation Weserübung (Weser Exercise). Due to the distances involved, support from both Luftwaffe air transport units as well as Kriegsmarine units was deemed essential for the success of the invasion. By late March, the date for the invasion was set to the next full moon, which would occur in early April. The invasion of Denmark and Norway would be the real baptism of fire for the Luftwaffe transport units.

The fact that large air, naval, and army forces were being gathered in northern Germany did not go unnoticed. During one RAF photo reconnaissance sortie against the Kiel area on 7 April, the airfields were seen to be brimming with Ju 52/3m transports.

On 9 April 1940, 571 Ju 52s were operational, with one *Transportgeschwader*, ten *Transportgruppen*, as well as one *Seefliegergruppe*. The latter would transport troops to the coastal town of Bergen.

Luftwaffe Ju 52/3m Units: 9 April 1940

Unit	Base	Total number of aircraft
Stabsschwarme	Ütersen	3
I:/KG z b V 1	Ütersen	53
II./KG z b V 1	Schleswig, Stade	53
III./KG z b V 1	Ütersen, Hagenow	53
IV./KG z b V 1	Hagenow	53
KGr z b V 101	Neumünster	53
KGr z b V 102	Oldenburg	53
KGr z b V 103	Schleswig	53
KGr z b V 104	Stade	53
KGr z b V 106	Ütersen	53
KGr z b V 107	Fuhlsbüttel	53
KGr z b V 108 *	Rantum, Hörnum	13

* Floatplane unit, which also had fourteen Heinkel He 59 floatplanes.

During the invasion of Denmark and Norway, more than 3,000 individual sorties were flown on 9 April, transporting 29,280 troops. Apart from troops, many tons of cargo, including petrol, was transported by the Ju 52/3ms. Denmark was quickly overwhelmed with a trio of Ju 52s from 8./KG b z b V 1 dropping paratroopers over Aalborg at 7 a.m. With the airfield secured, it was quickly pressed into use as an intermediate staging point for the Norwegian invasion force. No less than fifty-three Ju 52/3ms landed in quick succession, with ten suffering damage in landing accidents. One additional aeroplane was severely damaged when one of its engines caught fire. Nine more Ju 52/3ms dropped paratroopers over Storstrøm Bridge (*Storstrømsbroen*) ferry station near Copenhagen. With Danish airfields secured, transporting men and material to Norway continued apace.

Conquering Norway proved somewhat more difficult; the Norwegian defences showed determined opposition against the German invaders. A flotilla of Kriegsmarine vessels sailing towards Oslo was met by gun fire and torpedoes fired from Oscarsborg Fortress. The cruiser *Blücher* was sunk, with *Lützow* being damaged. Securing the Oslo Airport, Fornebu, (which had only been inaugurated the previous year) was a primary objective. Twenty-nine Ju 52/3ms of II./KG z b V 1 carrying paratroopers arrived over the Norwegian capital during the morning of 9 April. Met by poor weather and Norwegian Gladiators, one Ju 52/3m (W Nr 6570) was shot down, with the remaining aircraft being forced to return to Aalborg. It was an inauspicious beginning. However, twenty minutes behind the first wave of Ju 52/3ms were fifty-three aircraft of KGr z b V 103 carrying regular infantry. This formation headed towards Oslo Airport, Fornebu, despite a radio message ordering them to turn back (apparently, the unit commander *Hauptmann* Richard Wagner believed the radio message was a deception), arriving overhead at 9.05 a.m. By this time, the fog and low clouds had dissipated.

The first Ju 52/3m, in which Wagner was flying as a passenger, was severely hit during the attempted landing, with Wagner being killed. A few Messerschmitt Bf 110s managed to provide enough cover for the vulnerable transports. Chaos reigned as several Ju 52/3ms were damaged by Norwegian defensive fire, as well as suffering from crash landings. The transport aircraft were running out of fuel, landing wherever possible. With the airport being under German control, a radio message was relayed to the Headquarters of X. Fliegerkorps that the Ju 52/3ms had managed

Luftwaffe Ju 52/3ms at Oslo's Fornebu Airport in April 1940, with the sole remaining Junkers G 38 seen just prior to landing. (*Mats Averkvist*)

to land at Fornebu. Orders were immediately sent to the remaining transport units to take-off and head north. Within eight hours, another 159 Ju 52/3ms had landed at Fornebu.

Simultaneous with the fight for Fornebu, Ju 52/3ms of KG z b V 1 and KGr z b V 104 arrived over Stavanger. The first of 131 paratroopers jumped over Sola airfield at 9.20 a.m. The airfield was occupied with little difficulty, with more than 100 Ju 52/3ms of KGr z b V 104 and 106 bringing in reinforcements. At Bergen, a dozen Ju 52/3m floatplanes and Heinkel He 59s of KGr z b V 108 transported troops and supplies. Troops and supplies were continuously ferried from Germany to Norway by the Ju 52/3ms. On 13 April, two *Nachrichtenkommandos* and their equipment were airlifted to Fornebu and Trondheim Airport, Værnes.

During the airlift operation, some of the Ju 52/3ms crossed into neutral Sweden, with about ten being shot down by Swedish anti-aircraft guns. Three of these came down in Sweden, with others being observed to crash into the sea.

Following the arrival of British troops, it was decided to mount a paratroop attack against Dombås in central Norway during the evening of 14 April. Lacking maps and up-to-date information about the enemy's strength and positions, fifteen Ju 52/3ms of II./KGr z b V 1 attempted to drop their load of *Fallschirmjäger* over Dombås. Due to heavy ground fire and poor weather, the operation was a disaster. Several Ju 52/3ms were shot down or badly damaged.

The city of Narvik was captured by a small German force. However, supplying the forces and sending reinforcements to the strategically important Norwegian town was extremely difficult. On 13 April, thirteen Ju 52/3ms of KGr zbV 102 left Berlin Tempelhof. Aboard were a contingent of sixty troops from 2./A.R. 112 *Gebirgsbatterie*, as well as supplies, including four 120-mm guns. One aircraft had to turn back at Kiel due to engine trouble. Reaching Oslo with precious little fuel (the planned refuelling at Aalborg had to be cancelled), the dozen Ju 52/3ms were joined by a radio-equipped Ju 52/3m. The resupply flight to Narvik was a one-way mission, as the Ju 52/3ms did not have the necessary range to fly to Narvik and make a safe return flight to the German lines.

During the flight northwards, two of the aircraft had to put down at Gullesfjorden due to poor weather, and were attacked by three Norwegian Navy Heinkel He 115s the following day. The remaining

Above: Soon after its capture at Oslo's Fornebu Airport, the British Airways Ju 52/3m G-AFAP had its registration letters amended to G+AP. (*Mats Averkvist*)

Right: A former DNL Ju 52/3m taken over by the Luftwaffe in April 1940. The Norwegian civilian registration letters are still faintly visible. (*Swedish Aviation Historical Society*)

One of the Ju 52/3ms lost during the early stages of Operation Weserübung. (*Mats Averkvist*)

crews got as far as Lake Hartvikvannet just south of Narvik. Here, it was discovered that the ice had begun to thaw. Large puddles of water made it impossible to consider flying the Ju 52/3ms off the lake. As a result, the German aircraft were stuck. Using enforced local Norwegian labour, a primitive runway was constructed by packing snow. A call for assistance was made, with a trio of Ju 52/3ms eventually arriving over the lake early the next morning. However, instead of the expected fuel, food and a field gun were air-dropped, with the latter ending up with the gun barrel firmly stuck into the frozen ground. Subsequent Norwegian and British air attacks saw four of the Ju 52/3ms destroyed and others damaged. One of the Ju 52/3ms (with additional fuel drained from the other stranded aircraft) managed to lift off Lake Hartvikvannet on 17 April, only to stray across the border into Sweden. On 19 April, the Germans abandoned the stricken Ju 52/3ms, with Norwegian forces recapturing Lake Hartvikvannet. Three of the Ju 52/3ms were put back into flying condition, with empty fuel drums being attached to the aircraft. The intense effort to repair the former Luftwaffe transports was in vain with RAF Hurricanes of 46 Squadron strafing the Ju 52/3ms on 24 May. Before this took place, engines, instruments, and other equipment had been removed from several of the Ju 52/3ms.

All of the aircraft sank to the bottom of the lake when the ice thawed. German forces found and recovered the wrecks of two or three aircraft, leaving the others in their watery grave (in 1982, one of the Ju 52/3ms was brought up to the surface, with others being recovered during the following years). On 10 June, the last Norwegian forces surrendered to General Dietl. The exact number of Ju 52/3ms lost during the Norwegian campaign is difficult to establish, but was possibly as high as 120–130 aircraft.

Crete

In the event, the assault on Crete, codenamed Operation Mercury (*Unternehmen Merkur*), was postponed until 20 May. More than 20,000 German troops would participate in Operation Mercury, including 15,750 *Fallschirmjäger*. Ten *Transportgruppen* with 520 Ju 52/3ms as well as the I./Luftlandegeschwader 1 equipped with seventy-two DFS 230 assault gliders would carry the *Fallschirmjäger* into combat. The Ju 52/3m units were: I./KGr z b V 1, II./KGr z b V 1, KGr z b V 40, KGr z b V 60, KGr z b V 101, KGr z b V 102, KGr z b V 105, KGr z b V 106, I./KGr z b V 172, and II./KGr z b V 172. Though unknown at the time, the invasion of Crete was to be the final large-scale German air assault. Despite reaching its objective, the German *Generalleutnant* Kurt Student later said that Crete was 'the grave of the German paratroop force'.

At 5.30 a.m. on 20 May, the Ju 52/3ms began to take-off. The dusty airfields created severe visibility problems, with the dust rising to 1,000 metres. As a result, there were long intervals between take-offs. Nevertheless, the Ju 52/3m *Gruppen* assembled over Kythera Island and headed for Crete. The first Ju 52/3ms to leave Greece were those of I./LLG 1, towing DFS 230 assault gliders. Arriving over Tavronitis, many of the DFS 230s crashed fatally. Nevertheless, enough *Fallschirmjäger* survived to silence the anti-aircraft gun positions. At 8.20 a.m., the first Ju 52/3ms of the main formation arrived over Maleme airfield. However, several Ju 52/3ms of KGr z b V 102 were shot down, with one falling into several *Fallschirmjäger* who only moments before had jumped out of another Ju 52/3m. More than 150 Ju 52/3ms dropped their load of *Fallschirmjäger* west of Chania, with only two or three aircraft being lost. In all, no more than seven Ju 52/3ms of the initial wave were lost; however, upon returning to Greece, several aircraft were destroyed or damaged in landing accidents.

The arrival of the Ju 52/3ms in the air above Crete was later described in the *New Zealand Expeditionary Forces Times* (an armed forces newspaper/bulletin issued during the war), 18 January 1943 issue:

The screaming Junkers over the grey-green trees,
Their cargoes feathering to earth.
They might be wisps of white rose petal
Caught in the keen, compelling twist of fate
Faltering, aimless, in an aimless wind:
Confetti, white and dirty white,
Tossed out in scattered handfuls...
And one man idle, leans against the open hatch,
Through which the white horde poured
And watches Crete whine past below,
And in the mixed array of conquest
His hearing does not catch the rifle snap,
Sudden, faint his hands grasp deeply into nothingness,
And in bewildered agony
The dark soul drowns.
He struggles as the troop plane banks;
Unstruggling, falls in one slow turn—
The horror dream personified—
And the olives snatch him to their greenery.
Our vague ears do not catch the death—weak cry,
And someone blows the smoke shreds from his rifle mouth.[4]

The German HQ was unaware of the heavy fighting in Crete and that the *Fallschirmjäger* had failed to capture Maleme, Chania, or Souda Bay. Later in the day, a Ju 52/3m landed at Maleme only to be fired upon. The pilot immediately pushed the throttles forward, and managed to take-off and return to Athens—as a result, the second wave of Ju 52/3ms was delayed. Nevertheless, a second wave numbering twenty-four Ju 52/3ms arrived over the island at 4.15 p.m. At least seven aircraft were shot down by Australian troops, and many of the *Fallschirmjäger* were hit during their descent or immediately upon hitting the ground. A third wave of some 240 Ju 52/3ms arrived at low altitude over Heraklion airfield. Intense anti-aircraft fire accounted for about fifteen aircraft lost, which came down over or near the airfield, while others were observed ditching into the sea.

Subsequent air drops the following day suffered little in the way of anti-aircraft fire, with most gun positions having been knocked out. However, getting reinforcements and heavy guns into the landing zones saw a number of Ju 52/3ms performing crash landings at Maleme. Over eighty Ju 52/3ms were strewn across the airfield in various stages of disrepair. A number had made hard landings, wiping the undercarriage off, while others had burnt out.

With Maleme airfield in German hands, a continuous flow of Ju 52/3ms arrived, carrying troops, ammunition, and other supplies. On 23 May, RAF bombers attacked Maleme, claiming ten Ju 52/3ms destroyed. Six aircraft of KGr z b V 106 and I./LLG 1 were in fact destroyed. On 24 May, at least twenty-three Ju 52/3ms were lost at Maleme, the majority in landing accidents (including ten from KGr z b V 1, six from KGr z b V 102, and seven from I./LLG 1).

Further RAF raids in late May accounted for another seven Ju 52/3ms destroyed. By 27 May, however, the situation for the remaining British and Commonwealth troops in Crete had worsened considerably. An order to evacuate to Egypt was subsequently issued.

By 1 June, Crete was in German hands, but the cost had been horrendous. Of the 22,000 troops committed, 6,543 had been killed, gone missing, or seriously wounded. Among these were 3,764 *Fallschirmjäger*. Indeed, the *Fallschirmjäger* would never again be committed to air drops on such a large scale. No less than 117 Ju 52/3ms had been shot down or written-off in crashes, while another 125 had been damaged. Other sources speak of 271 Ju 52/3ms being lost to all causes. While most of these were shot down by flak or RAF fighters, several were destroyed at Maleme during RAF bomb raids. The extraordinarily high losses of aircraft and crews resulted in two *Transportgruppen*, KGr z b V 40 and 60, being disbanded, with their aircraft being transferred to the units that had suffered the highest losses.

The aftermath of Operation Mercury with numerous Ju 52/3ms in various stages of disrepair. Many of the aircraft were literally flown into the ground in a desperate attempt to reinforce the German troops. (*Author's Collection*)

8
Further Development

Although the basic Ju 52/3m airframe remained essentially the same during the entire production run, it was soon realised that both DLH and the Luftwaffe required larger and more capable transport aircraft. Two types of transport aircraft entered limited production during the war as the Junkers Ju 252 and Ju 352 respectively. The thoroughly tested trimotor configuration of the earlier Ju 52/3m was retained for both the Ju 252 and Ju 352.

Junkers Ju 252

Already in 1938, Junkers initiated the design of an airliner development of the Ju 52/3m. Working in close cooperation with Lufthansa, the new airliner was to feature the thoroughly tested three-engine layout. However, the corrugated aluminium construction of the earlier Ju 52/3m was to be replaced by a smooth metal skinning. The new airliner was also to feature a fully retractable undercarriage. Several design studies were put forward, with the Junkers EF 77 design being accepted in December 1938—the EF 77 could carry twenty-one passengers. Changing DLH requirements called for an aeroplane with better performance, which saw the EF 77 remain on the drawing board. The new airliner was designated Ju 252, and the first prototype, the Ju 252 V1, first flew in late 1941, registered as D-ADCC. Within the next couple of months, the second and third prototypes joined the flight test programme. Meanwhile, a batch of twenty-five Ju 252s had been laid down at Bernburg in anticipation of a DLH order. However, no Ju 252s would reach DLH due to the Luftwaffe's desperate need for military transport aircraft. The original order for twenty-five was subsequently reduced to fifteen. Some component production was performed by Fokker, with sixty-four oil tanks and sixty 'utility engines' being built.

The first Ju 252 taken on charge by Luftwaffe was the fifth prototype, the Ju 252 V 5. It was delivered to *Lufttransport Staffel* 290 (LTS 290) in January 1943. It did not last long, being destroyed in an Allied air raid at Grosseto, Italy, in March 1943. Soon afterwards, LTS 290 was redesignated *Transport Staffel* 5, receiving several Ju 252s. The third prototype, Ju 252 V 3, served as a troop transport in August 1943, while two Ju 252s served with I./KG 200. One, coded T9+SK (W Nr 0008), was destroyed in a crash at Petroşani, Romania, on 26 June 1944.

No intact examples of the Ju 252 were found by the Allies.

Although the Ju 252 showed promise, only fifteen were completed. Planned variants of the Ju 252 included the Ju 252B with increased armament, the Ju 252C, which was to have been powered by three Bramo 323 engines, and the Ju 252D floatplane.

Junkers Ju 252 Technical Data and Performance

Engines:	3 × 1,340-hp Junkers Jumo 211F
Armament:	One 13-mm MG 131 and 2 × 7.9-mm MG 15
Wingspan:	34.09 m (111 feet, 10⅛ inches)
Length:	25.10 m (82 feet, 4¼ inches)
Height:	5.75 m (18 feet, 10⅓ inches)
Wing area:	122.6 m^2 (1,319.652 square feet)
Empty weight:	13,100 kg (28,880 lb)
Weight (loaded):	22,480 kg (49,560 lb)
Maximum speed:	438 kph (272 mph)
Range with maximum payload:	3,980 km (2,473 miles)
Service ceiling:	6,300 m (20,670 feet)

Junkers Ju 352 Herkules

During the spring of 1942, the design team at Dessau received a request from the RLM for a development of the basic Ju 252. The new aircraft was to be constructed using as much non-strategic materials as possible. Aerodynamically, the Ju 352 was essentially similar to the earlier Ju 252; however, due to the use of wood, it was an entirely new aeroplane.

The wing was built entirely out of wood, with the fuselage being of mixed construction. The forward and rear sections were built up from duraluminium-stressed skin, with the fabric-covered centre fuselage constructed out of steel tubing. The Junkers Jumo 211F engines were replaced with three 1,000-hp BMW-Bramo 323 radial engines.

Junkers unofficially applied the name Herkules to the new transport. The rear fuselage featured a loading ramp, named *Trapoklappe*, which was hydraulically operated. It was possible to drive vehicles up the ramp, however, this was forbidden as the risk of damaging the internal structure was deemed too great. Instead, vehicles and heavy equipment were manually winched aboard.

Due to the fact that the Junkers factories lacked the necessary space, the Ju 352 programme was transferred to Fritzlar. Here, the Ju 352V1 flew on 1 October 1943, with the second prototype following a couple of weeks later. Before any of the prototypes had been flown, a pre-production batch of ten Ju 352A-0s was ordered by the Luftwaffe. The test flights revealed the Ju 352 to be a vast improvement over the Ju 52/3m. The pre-production aircraft had all been delivered by February 1944, and the first series production Ju 352A-1 was delivered in April, but only thirty-three (thirty-one according to some sources) of this variant were built before production ended in September.

Four Ju 352s were used by KG 200, including T9+AB and T9+CB with Stab KG 200, A3+HN of 5./KG 200, and A3+AB of I./KG 200. Having an adequate range and good cargo-carrying capability, the Ju 352s of KG 200 were used for various cloak-and-dagger operations, including delivering agents behind Allied lines. Most of the Ju 352s were supplied to the *Grossraum-Transportgruppe* (Large Capacity Transport Wing) based at Tutow. As late as 25 April 1945, twenty-three Ju 352s were still operational with this unit; however, complete Allied air supremacy and lack of fuel severely restricted operations.

One Ju 352A of the *Grossraum-Transportgruppe* (GG+NX) ended up in a farmer's field west of Thisted, Denmark, on 4 May 1945. The pilot, Artur Kürth, and his Danish fiancée, Aase Krogh, had attempted to reach Frederikshavn, when engine trouble and lack of fuel forced Kürth to belly-land. The huge Junkers remained *in situ* for some weeks until dismantled by Zonen (a Danish rescue organisation) and transported to Aalborg for disposal. A few Ju 352s were found in May 1945 at Fliegerhorst Schleswig-Land (Jagel), Germany.

Three Ju 352s were briefly flown by the RAE. The first of these was the final pre-series aircraft, Ju 352A-0 W Nr 100010 (KT+VJ), which was surrendered at Flensburg. It was air-tested on 22 June 1945 and flown to Farnborough the following day. The Ju 352A-0 was assigned the serial number AM 10, later changed to VP550. The transport was used extensively for trips to the continent to collect material deemed worthy of examination by RAE, with several flights being made to an RLM research centre at Völkenrode, and the final flight took place on 28 June 1946. The Ju 352 was scrapped at RAE Farnborough, being noted in the scrap area on 15 December 1946.

The other two RAE Ju 352s were both surrendered at Eggebek. They had belonged to V./TG 4, being coded G6+RX and G6+SX. The former became AM 109, being ferried from Schleswig to Farnborough on 31 August 1945. Only one return trip to the continent was made, and this was to Schleswig on 4 September. It seems to have been withdrawn from use soon afterwards, being noted in the Farnborough scrap area on 15 September 1946.

Ju 352A (W Nr unknown) surrendered at Eggebek, code G6+SX, unit V./TG 4 and became AM 110. Either AM 109 or AM 110 suffered engine problems during the ferry flight to Farnborough, resulting in a forced-landing at Knokke, Belgium. The Ju 352 was reportedly offered to the Belgian government, which declined, and the Ju 352 was reduced to scrap shortly afterwards.

One Ju 352 was flown by the Czech Air Force, being designated D-352. One flight took the Ju 352 to Moscow, and in its cargo hold was the first Tatra car built, which was presented as a gift to Stalin. The Ju 352 was subsequently registered as OK-JUE, and its fate is unknown. One Ju 352 was also used in the USSR as an engine test bed.

Junkers Ju 352 Technical Data and Performance

Engines:	3 × 1,200 hp-BMW-Bramo 323R-2
Armament:	One 13-mm MG 131 and 2 × 7.9-mm MG 15
Wingspan:	34.21 m (112 feet, 2¾ inches)
Length:	24.60 m (80 feet, 8½ inches)
Height:	5.75 m (18 feet, 10⅓ inches)
Wing area:	128.19 m^2 (1,379.93 square feet)
Empty weight:	12,769 kg (28,150 lb)
Weight (loaded):	19,595 kg (43,200 lb)
Maximum speed:	370 kph (230 mph)
Range with maximum payload:	2,995 km (1,860 miles)
Service ceiling:	6,000 m (19,685 feet)

9
Into Luftwaffe Service

The German Luftwaffe was the largest operator of the Ju 52/3m. It was the mainstay of the *Transportverbände* (Air Transport units), serving on every front, from the freezing cold of the Arctic to the suffocating heat of North Africa. Ju 52/3ms were highly appreciated by the German troops. Several nicknames were bestowed on the Ju 52/3m, including *Eisernes Annie* (Iron Annie) and *Alte Tante Ju* (Old Lady Ju). Apart from the standard three-engined Ju 52/3m, at least one single-engined Ju 52 was used by the Luftwaffe as a target tug.

Although the German air arm had been largely disbanded following the end of the First World War, plans formulated in June 1927 included the purchase of 247 aircraft for a future military air arm. Twenty of these were to be Junkers G 24 *Behelfsbomber* (auxiliary bombers). In this context, mention must be made of the Junkers K 30, a bomber variant developed in 1925 from the G 24 for delivery to the Soviet air arm. During the autumn of 1928, one Junkers G 24 was tested at Lipetsk, USSR, as a bomber. Several different military aeroplanes were secretly tested at Lipetsk, with many German pilots receiving military pilot training following the signing of an agreement in December 1923. In 1928, the Director of Junkers, Gotthard Sachsenberg, submitted the first of two memorandums, in which he suggested that a fleet of 100 auxiliary bombers be acquired. Such a large order would inject capital into the German aviation industry (particularly Junkers) as well as making an impression on foreign powers. Unsurprisingly, Sachenberg proposed that such a bomber should be based on the existing Junkers G 24. The *Reichswehrministerium* were not particularly impressed, suspecting that Sachenberg was angling for an economic infusion from the government. With Junkers being heavily involved in a wide diversity of aeronautical experiments, the company was losing money. Additionally, Junkers' military aeroplanes had a fairly poor reputation.

On 24 February 1932, the name Luftwaffe was accepted for the re-established (albeit still secret) German air arm. It was planned to acquire Junkers G 24s from Lufthansa for conversion as night bombers. In the event, the more modern (and in production) Junkers Ju 52/3m would be seen as pivotal in rebuilding a German bomber force. In 1932, Erhard Milch proposed that a large quantity of Ju 52 *Behelfsbomber* (auxiliary bombers) should be acquired (similar to Sachenberg's proposals of 1928 and 1929) at the cost of 4 million Reichsmark, (RM), which was ultimately rejected.[1]

Although it was agreed that the performance of the Ju 52/3m in the bomber role was marginal, it would go on to serve as a *Behelfsbomber*. As such, Luftwaffe bomber crews began honing their skills in the slow, lumbering Ju 52/3m.

In 1933, a Ju 52 ce, registered SE-ADM (W Nr 4004), was tested in Norway as a torpedo bomber. The Norwegian-designed torpedo had an all-up weight of 610 kg, of which 150 kg constituted the war load. The trials involved dropping the torpedoes from up to 80 metres altitude. In all, sixty test drops were made between 29 May and 2 June and also between 4 September and 17 October. German personnel from the Test Centre at Travemünde (E-Stelle Travemünde) and the Torpedo Trials Centre participated in the trials. This torpedo (with an increased 175-kg war load) was adapted by the Kriegsmarine and was designated F5.

On 10 April 1933, just over three months after Hitler's rise to power, a plan entitled 'Preparations for the Creation of a Peace-Time Air Force' was put forward. In all, nine bomber squadrons were to be formed, and the plan was revised on several occasions. The Ju 52/3m was designated as a *Behelfs-Kampfflugzeug* (auxiliary bomber aircraft). Beginning in 1933, Ju 52/3ms were delivered to the still-secret Luftwaffe, serving with a number of Kampfgruppen. On 1 March 1934, twenty-four Ju 52/3m were on charge with *Behelfs-Kampfgeschwader* (auxiliary bomb group). The first Luftwaffe unit to receive Ju 52/3ms was *Bombengeschwader* 1 (first bomb wing), based at Berlin Tempelhof. Soon afterwards, *Bombengeschwader* 1 was divided into two units, *Behelfs-Kampfgeschwader* 172 (based at Berlin-Tutow) and 274 (based at Fassberg), both operating Ju 52/3m g3e and g4e variants designed from the outset as auxiliary bombers. The maximum bombload carried by the g3e was 1,500 kg (3,307 lb) carried in three cells each containing a vertically stowed 500-kg bomb. Depending on distance and type of target, the load of bombs carried could vary between 1-kg incendiary bombs (a maximum of 864 could be carried), ninety-six C10 bombs, twenty-four C50 bombs, six C250 bombs, or three C500 bombs. The bomb sight, a Goertz Fl 219b (essentially the same as used on the First World War-era Gotha bombers), had a number of shortcomings, which made actually hitting the designated target difficult.[2]

The defensive armament consisted of two MG 15 machine guns, one in an open dorsal gun position and one in a manually retractable 'dustbin' (nicknamed '*Topf*' or 'Pot') in the lower fuselage—the *Topf* was also an integral part of the later Ju 86 twin-engined bomber. The latter gunner also doubled as bomb aimer. Although the *Topf* provided an excellent view for the bomb aimer, as well as an excellent field of fire for the gunner, it made the aircraft more difficult to manoeuvre when extended. The crew of the Ju 52/3m auxiliary bomber consisted of four: pilot,

Ju 52 SE-ADM seen during torpedo trials. (*Mikael Forslund*)

radio operator, dorsal gunner, and ventral gunner/bomb aimer, with the latter being designated as commander.

Another unit of great importance for the development of the future Luftwaffe bomber force was the *Hanseatische Fliegerschule e.V.* Formed at Fassberg, this was in essence a conversion unit, equipped with both Ju 52/3ms and Dornier Do 11Cs.

In 1934, Junkers received an order for 1,200 Ju 52/3ms, the largest order at the time. These aircraft were all destined for the future Luftwaffe. Although designed as a transport aircraft, the Luftwaffe Ju 52/3ms could be converted as bombers for the *Behelfs-Kampfgeschwader*.

On 1 March 1935, the existence of the new-born German air arm, Luftwaffe, was revealed. A number of *Luftkreiskommandos* (literally regional air commands) were established. The pair of *Behelfs-Kampfgeschwader* were redesignated as Kampfgruppe I./KG 252 (based at Berlin-Tutow) and I./KG 154 (based at Fassberg), forming part of *Luftkreiskommando* II and IV respectively.

Seven months later, on 1 October, no less than seven *Kampfgeschwadern* had been formed. Out of these, I./KG 154, I./KG 252, II./KG 352, I./KG 455, I./KG 553, and I./KG 653 operated a total of seven *Staffel* of Ju 52/3ms.

It was intended that the Ju 52/3ms would be replaced by the Dornier Do 11C, which was specifically designed from the outset as a bomber. Compared to the rugged and easy-to-fly Ju 52/3m, the Do 11C proved to be more lethal to its crews than anything else. In fact, the crews flying the Do 11C quickly named it *Fliegender Sarg* (Flying coffin). Only three Do 11Cs were in service on 1 March 1934. In the event, the enormous difficulties with the Do 11 meant that an improved variant, the Do 23, entered production. Nevertheless, the Do 23 only saw limited service as a bomber, with most quickly being relegated to various second-line duties.

By the end of 1936, no less than eight *Kampfgruppen* (each with thirty-six aircraft) were equipped with the Ju 52/3mge and g3e variants. Although considered obsolete as a combat aircraft, the Ju 52/3m was at the time the Luftwaffe's most numerous bomber. One aspect of the Ju 52/3m is that it was an excellent aircraft for the future bomber crews to cut their teeth on.

With dedicated bombers such as the Dornier Do 17, Junkers Ju 86, and Heinkel He 111 joining the Luftwaffe, the majority of the Ju 52/3ms were transferred to the role of advanced trainer, serving with a large number of *C-Schulen* (a *C-Schule* was where the bomber and transport pilots received their multi-engine training before transfer to a combat unit). The Ju 52/3m was rugged and a wonder of simplicity, being very forgiving to novice pilots. One Luftwaffe pilot, Werner Utter, later recalled a blind-flying exercise, during which he attempted to peer outside the screen, while simultaneously checking the instruments. Utter confused the ground lights with a bright star, with the Ju 52/3m performing a cross between a loop and a barrel roll. During precision landing exercises, which involved turning off the engines at an altitude of 1,000 metres, a colleague of Utter's always ended up undershooting or overshooting the runway. The solution was to continue the glide with crossed rudders, and, if discovering that he would overshoot, order the doors to be opened. The doors functioned as *ad hoc* air brakes, with the Ju 52/3m sinking like a stone; however, before pulling up, it was very important to close the doors.

In March 1935, German forces marched into the Rheinland (Rhineland), which had been under French occupation since the end of the First World War. Although not a shot was fired, it was feared that the French would send bombers to attack German positions. Reportedly, Ju 52/3m bombers of KG 257 were placed on readiness to bomb Paris in retaliation if the French decided to attack the German troops.[3]

In October 1937, IV./KG 152 'Hindenburg' was transferred from the bomber role to that of dedicated air transport, with the unit being redesignated as '*Kampfgruppe bis zur besonderes Verwendung*' 1 (KGr b z b V 1)—'combat unit for special employment'.[4]

The role envisaged for the unit was that of transporting and dropping paratroopers over the combat zone, with a secondary role as a bomber (Luftwaffe Ju 52/3ms would eventually be used as bombers on at least two occasions during the Second World War). Based at Fürstenwalde/Spree, KGr b z V 1 initially had thirty-nine Ju 52/3ms. On 1 August 1938, KGr b z b V 2 was formed.

10
Other Duties

Apart from its use as a cargo and paratrooper transport, the Ju 52/3m was operated by the Luftwaffe in a wide variety of roles. Originally formed in the 1920s as a resource for correct and up-to-date weather information for civilian airliners, the *Wetterdienst* (weather service) was an important Luftwaffe function. When war broke out in September 1939, there were five Wekusta (*Wettererkundungsstaffeln*, weather reconnaissance squadrons), although several more Wekustas were subsequently formed. Wekusta 1, 26, and 51 are known to have included Ju 52/3ms among their aircraft inventories. Casualty evacuation was usually performed by the *Sanitätsflugbereitschaften* (air ambulance squadrons). In all, fourteen *Sanitätsflugbereitschaften* were formed, with the majority being established in the spring of 1941, prior to Operation Barbarossa.

One important aspect of the Ju 52/3m was its use as an advanced trainer. With dedicated bombers such as the Do 17, He 111, and Ju 86 entering Luftwaffe service, the Ju 52/3m auxiliary bombers were largely transferred to other duties. The future Luftwaffe bomber or transport pilot would have to pass the C2 level. This was achieved at a *Flugzeugführerschule* C (literally C-level pilot training school). However, the majority of the pilots passing through the *C-Schule* were predestined as bomber pilots. The typical *C-Schule* aircraft had two or more engines, an all-up weight of more than 5,500 kg as well as room for six persons. Many Ju 52/3ms served with the various *C-Schule*, with many of the instructors at the *C-Schule* being former Lufthansa pilots who had vast experience of flying Ju 52/3m's through all kinds of weather. However, the requirements for specialised transport pilot training were not recognised until mid-1942, when KGr b z V 300 was formed as a dedicated training unit. In May 1943, KGr b z V 300 was reorganised as a specialised *Ergänzungsgruppe* (replacement group). By 1943, the need to replace the huge losses of pilots and crews necessitated a revised and considerably shorter training schedule. Many of the *C-Schule* were reorganised as *A-* or *B-Schule*. The quality of flight training was considerably reduced as well, with many of the most experienced instructors, including many Ju 52/3m pilots, being transferred to combat units. With the training units being continuously drained of both instructors and aircraft, the net result was, in the long run, disastrous for the Luftwaffe's *Transportverbände*.

Luftwaffe FFS *C-Schule*[1]

Unit designation	Established	Base/s	Remarks
FFS C 1	01.1940	Pyritz (Pyrzyce), Sorau (Żary)	Closed 09.1942
FFS C 1	03.1943	Schweinfurt, Bayreuth-Bindlach	Redesignated FFS A 7 in 07.1944
FFS C 2	04.1939	Neuruppin, Schwerin	Redesignated FFS B 2 in 10.1943
FFS C 3	01.1940	Alt-Lönnewitz, Sprottau (Szprotawa)	Redesignated FFS B 3 in 10.1943
FFS C 4	01.1940	Perleberg, Sprottau, Barth	Redesignated FFS B 4 in 10.1943
FFS C 5	01.1940	Ludwigslust, Neubrandenburg	Redesignated FFS B 5 in 01.1943
FFS C 6	01.1940	Burg bei Magdeburg, Stade, Kolberg	Redesignated FFS B 6 in 10.1943
FFS C 7	01.1940	Celle, Finsterwalde, Clermont-Ferrand, Gablingen	Redesignated FFS A 5 in 07.1944
FFS C 8	01.1940	Fürth-Atzenhof, Wiener-Neustadt, Eisenstadt Trausdorf, Parndorf	Redesignated FFS B 8 in 10.1943
FFS C 9	01.1940	Roth-Kiliansdorf, Altenburg, Pretzsch	Redesignated FFS B 9 in 10.1943
FFS C 10	01.1940	Landau, Fürstenwalde	Redesignated FFS B 10 in 10.1943
FFS C 11	01.1940	Zeltweg/Steiermark (Styria)	Redesignated FFS B 11 in 10.1943
FFS C 12	01.1940	Wesndorf, Prague-Ruzyne	Redesignated BFS 3 in 07.1942
FFS C 12	02.1943	Olmütz (Olomouc)	Redesignated FFS B 12 in 10.1943
FFS C 13	01.1940	Rosenborn, Nancy, Toul, Delme, Roth-Kiliansdorf, Öttingen	Redesignated JG 111 in 06.1944
FFS C 14	12.1939	Ohlau (Oława), Prague-Gbell	Redesignated FFS B 14 in 10.1943
FFS C 15	01.1940	Lüben, Gablingen, Bourges, Steinamanger (Szombathely)	Redesignated FFS B 15 in 10.1943
FFS C 16	05.1940	Burg/Madgeburg, Weissenwarthe, Sachau	Closed 04.1945
FFS C 17	01.1941	Pütnitz, Greifswald, Bug, Güstrow, Barth	Originally FFS (See) 2. Closed 09.1944
FFS C 18	09.1941	Lüben, Schönfeld-Seifersdorf	Redesignated FFS B 18 in 10.1943
FFS C 19	09.1941	Ohlau (Oława), Gleiwitz, Woisselsdorf	Redesignated FFS B 19 in 10.1943
FFS C 20	10.1941	Rosenberg, Krosno, Tonndorf, Bednary	Redesignated FFS A 6 in 07.1944
FFS C 21	09.1941	Bialystok, Rosken, Königsberg, Thorn, Hannover	Closed 09.1944
FFS C 22	09.1941	Wien-Aspern, Oels (Oleśnica), Liegnitz (Legnica)	Redesignated FFS B 22 in 10.1943

Mausi Ju 52/3ms

With the RAF sowing magnetic mines along the coastlines of occupied Europe, the requirement arose of disposing of this threat. Different types of aircraft, including Blohm & Voss Bv 138 and Dornier Do 24 flying boats, were converted for anti-mine operations. Over seventy Ju 52/3ms were also converted, which included fitting a duraluminium Gauss ring, 14.6 metres in diameter beneath the fuselage. The ring was activated by means of a 270-hp eight-cylinder Mercedes Nürburg engine fitted inside the forward fuselage. Engine cooling was achieved through a radiator fitted in front of the undercarriage, and, to allow further air cooling, there were two inlets fitted to the fuselage—the additional gear weighed 1,136 kg. These aircraft were designated Ju 52/3m MS (*Minensuch* or minesweeper), with their crews referring to them as Mausi. The first Ju 52/3m MSs became operational with the *Minensuchgruppe* in November 1943. Aircraft were regularly deployed to the Baltic Sea, the French and German North Sea, and Atlantic coasts as well as the Adriatic Sea. Additionally, six aircraft were deployed to Budapest in April 1944 to clear the River Danube of mines dropped by the RAF. Detonating magnetic mines was a highly dangerous task, which involved flying at extremely low altitudes—usually between 10 and 30 metres. When a mine exploded, the aircraft ran the risk of sustaining damage from the shockwave and flying shrapnel. The minesweeping Ju 52/3ms always operated in pairs, being unofficially known as *Ehepaar* (married couple). The large ring was referred to as a *verlobungsring* (engagement ring).

In December 1943, two Mausi Ju 52/3ms (W Nr 3383 and 3400) were shot down by RAF fighters near Groix, France. Other Mausi losses occurred on 14 April 1944, when RAF Mosquitoes encountered a formation of Ju 52/3ms off the Danish coast. Several Mausi Ju 52/3ms were surrendered in Denmark and northern Germany. It would appear that even if some interest was evinced in the anti-magnetic gear, most Mausi Ju 52/3ms were reduced to spares or scrapped, the additional work to render them usable for civilian use being deemed excessive.

Ju 52/3m 'Wasser'

A number of Ju 52/3ms flew as floatplanes, particularly in Norway and the Aegean Sea. During the invasion of Norway, float-equipped Ju 52/3ms of KGr z b V 108 carried troops and supplies to places where conventional undercarriage aircraft were unable to operate. Later on, such Ju 52/3ms were used to resupply isolated German garrisons in the Aegean Sea. Among the units to operate the Ju 52/3m as floatplanes included *Seetransportstaffel* 1 (formed at Phaleron, Athens, in October 1943), *Seetransportstaffel* 2 (formed at Hommelvik, Norway, in October 1943), and *Seetransportstaffel* 3 (formed at Hommelvik, Norway, in October 1943). The latter unit was transferred to Greece in December 1943, by which time it had a full complement of twelve aircraft.

A fine study of a 'Mausi' Ju 52/3m. (*Author's Collection*)

Small numbers of Ju 52/3m W floatplanes were used for the supply of isolated garrisons, primarily in Norway and the Aegean Sea. (*Author's Collection*)

11
Blitzkrieg

On the Western Front, the frontline remained essentially the same for more than eight months. The apparent lack of action made some observers dryly referring to it as the 'Phoney War'. On 7 April 1940, a Ju 52/3m, CN+DT, was shot down by French M.S.406 fighters, crashing in flames near Maricourt. The Ju 52 belonged to the *Flieger Kompanie Des Luftnachrichten Regiment* ObdL (or 9. *Flieger Versuchskompanie*), being equipped with radio gear designed for detecting information on the French navigational system. The Ju 52/3m had most likely seen use in Spain, as French investigators found stencilling in Spanish inside the wreck.

On 10 May 1940, the Phoney War changed into Blitzkrieg when German forces invaded the Low Countries and France. Although the invading Germans were met with stiff resistance, Belgium and the Netherlands were soon overrun. Operations over Belgium on 10 May included two major air assaults. The first of these was the capture of the Belgian fort Eben-Emael, which was successfully completed using forty-two Ju 52/3ms of III./ KGr b z V 1 towing DFS 230 gliders. This was the first occasion that assault gliders had been used in combat. In order to create confusion, 120 dummy parachutists had been dropped beforehand. The German *Sturmabteilung* (assault battalion), led by *Hauptmann* Walther Koch, had been specially trained for this particular mission. By the following morning, the German troops had cleared the area of the last Belgian defenders.

The second operation, known as Operation Niwi (after the towns Nives and Witry-lès-Reims near the Luxembourg border), involved transporting two companies of soldiers who were to secure Nives and Witry-lès-Reims and the adjacent roads and bridges ahead of the advancing Panzers. With the majority of the Luftwaffe's Ju 52/3ms being required to support operations over the Netherlands, very few were available for Operation Niwi. In fact, a few Ju 52/3ms were available for Operation Niwi, but nowhere near as many that would have been needed. In their place, light Fiesler Fi 156 Storch spotter aeroplanes were to carry the troops. Apart from the pilot, the Fi 156s could only carry two soldiers. Although the Germans managed to secure both Nives and Witry-lès-Reims, twenty-two of the Storches were lost, as were two or three Ju 52/3ms.

Operations Over the Netherlands

At 4 a.m. on 10 May, Ju 52/3ms of KG b z b V 1 took off heading for the Netherlands. Their objective was to secure the airfields at Waalhaven, Valkenburg, Ypenburg, and Ockenburg, as well as the bridges at Moerdijk and Dordrecht.

Fallschirmjäger descending over Waalhaven, the Netherlands, on 10 May 1940. (*Thijs Postma*)

Ju 52/3ms destroyed at Valkenburg. (*Thijs Postma*)

The invasion did not evolve as planned, with the German forces facing stiff Dutch resistance. The *Fallschirmjäger* dropped over Waalhaven had not managed to conquer the airfield when the first Ju 52/3ms of KG b z b V 2 arrived overhead. When the Ju 52/3ms began to land, several ran into mines, destroying or damaging a large number of aircraft.

Although Ju 52/3ms managed to land at Valkenburg, the heavy aircraft sank into the soft surface. Dutch forces managed to recapture Valkenburg soon afterwards. At Ypenburg, anti-aircraft artillery accounted for eleven out of thirteen Ju 52/3ms of the first wave. With waves of Ju 52/3ms following, their crews had to set down wherever possible. Hardest hit was KGr b z b V 9, which lost no less than thirty-nine aircraft.

Nevertheless, by mid-May, the success of the German ground forces made air supply operations largely redundant. As a result, all Ju 52/3m units, except KG b z b V 1, were withdrawn, with the aircraft being returned to the *C-Schule*.

During the invasion of the Low Countries, hundreds of Ju 52/3ms had been destroyed or damaged—in total, 167 Ju 52/3ms were written-off. Workers from the Dutch aircraft manufacturer Fokker were recruited to recover Ju 52/3m wrecks around the Netherlands. Those Ju 52/3ms that were deemed repairable were transported to the Fokker works while many other airframes were picked clean of usable parts. An RAF raid during the night of 7–8 October 1940 resulted in both the building and some of the Ju 52/3ms being damaged.

A burnt-out Ju 52/3m lies wrecked on a Dutch beach in May 1940. (*Thijs Postma*)

12
Operations

Operation Sea Lion

With France defeated, Britain remained alone in the fight against Nazi Germany. With the Battle of Britain beginning in earnest in late June 1940, preparations went ahead for the invasion of Britain, which was codenamed Operation Sea Lion (*Unternehmen Seelöwe*). Potential landing areas across southern England were littered with obstacles, including non-airworthy and obsolete aeroplanes. Luftwaffe raids against radar stations and RAF airfields continued until late August, when the Luftwaffe bombers were redirected against London and other cities. The RAF pilots were flying several sorties per day—flying to near exhaustion. It was the effort of 'The Few' that saved Britain from invasion, and it should also be remembered that, throughout the summer, RAF bomb raids against French harbours accounted for the destruction of many naval vessels. In the event, Operation Sea Lion was cancelled on 17 September, with Hitler believing that Britain would be subjugated through Kriegsmarine U-boats and other naval vessels effectively blocking the sea lanes, thus forcing the country into surrendering to Germany. The majority of the Ju 52/3ms were returned to the various *C-Schule*; however, it was not long before they were returned to the transport units.

On 7 September 1940, a specialised Luftwaffe Signals Unit was formed at Zwolle, the Netherlands. Led by *Leutnant* Hermann Diehl, the unit attempted to establish tactics in locating RAF night bombers. Height-locating gear was fitted to Freya radar stations, and, in order to simulate a RAF bomber, a Ju 52/3m was used.

In the early hours of 28 October November 1940, the Italian dictator Benito Mussolini initiated an invasion of Greece. Having numerical superiority, it was expected that Greece would fall to the Italian military might. However, the Italian invasion force soon ran into trouble, with the Greeks fiercely fighting back. Mussolini pleaded for Luftwaffe assistance in airlifting reinforcements and supplies from Italy to Albania, used as a springboard for operations against Greece. Hitler conceded, and, on 10 December, fifty-three Ju 52/3ms of III./KG z b V 1 (based at Wesendorf and commanded by *Oberst* Rudolf Starke) began to arrive at airfields around Foggia. The detachment lasted until the end of February, by which time 1,665 troop-carrying sorties (transporting some 30,000 troops) and 2,363 supply sorties (carrying 4,700 tons of material) had been made. Some 10,000 wounded and sick Italian soldiers were brought back.[1] Not a single Ju 52/3m was lost during this operation. (Already on 4 November, Hitler saw the potential need to despatch German forces to assist their Italian allies. The day before, the first RAF contingent had arrived in Greece. Although plans for the invasion of

Luftwaffe Ju 52/3ms seen at the Fokker works following an RAF raid in early October 1940. (*Thijs Postma*)

the USSR were foremost in the Nazi dictator's mind, having British forces operating in Greece and threatening the southern flank resulted in plans for an invasion of Greece being put forward.)

North Africa: The Initial Phase

Although British and Italian forces had been fighting in North Africa since 10 June 1940, the Luftwaffe did not make an appearance until mid-January 1941. Units of the Luftwaffe's X. Fliegerkorps, based in Sicily, were then transferred to Tripolitania. Ju 52/3ms flew numerous supply sorties from Germany to North Africa in support of Luftwaffe operations. The first recorded loss of a Ju 52/3m during the North African campaign occurred on 10 February, when an aircraft of III./LN Regt 40 (W Nr 6835, VB+VC) was forced down near Mellieħa due to poor weather, suffering 80 per cent damage.

On 20 April, Fairey Fulmars claimed four Ju 52/3ms shot down out of a formation of five. The Ju 52/3ms belonged to KG z b V 9, and were conducting a supply flight to Africa. According to Luftwaffe records, only two Ju 52/3ms were lost, flown by *Oberfeldwebel* Josef Kastl and *Feldwebel* Walter Heyer. One Fulmar was hit by return fire, with the crew being killed.

On 11 November 1941, Operation Crusader began. Several Ju 52/3ms fell victim to strafing attacks, with others being shot down.

A tragic incident occurred on 9 December, when T.Col. Bruno Brambilla intercepted a pair of multi-engined aeroplanes over Timimi. Flying a Macchi C.202, Brambilla incorrectly identified them as RAF Wellingtons, and shot them down. The aeroplanes were, in fact, Ju 52/3ms carrying ground crews and spares for JG 27. *Oberfeldwebel* Schultz of II./JG 27 quickly took off and gave chase to the Italian fighter. Schultz managed to score hits in the tail section of Brambilla's fighter. However, Brambilla managed to escape, eventually landing back at Martuba.[2] Six ground crew of III./JG 27 died, with another four being injured.

By the end of December 1941, the wrecks of hundreds of Luftwaffe and *Regia Aeronautica* aircraft littered the airfields of North Africa, having been destroyed or abandoned during the retreat. At Derna, eighteen destroyed Ju 52/3ms were found; another two were located at El Tamimi, while twelve damaged Ju 52/3ms were counted at Benina.

Invasion of Yugoslavia and Greece

In January 1941, three new *Transportgruppen* were formed, KGr b z b V 40, KGr b z b V 50, and KGr b z b V 60. With plans for the invasion of Greece already well underway, Yugoslavia remained a problem. On 1 March 1941, a Ju 52/3m of IV./KG z b V 1 landed at Skopje, Yugoslavia, due to navigational error, and the crew and their aeroplane were interned. At the same time, Luftwaffe units were deployed to Bulgaria in anticipation of the invasion of Greece. On 17 March 1941, the Yugoslav government signed the Tripartite Pact, thus formally becoming an ally of Germany and Italy. However, a few days later, a *coup d'état* by senior officers made the pact null and void. As a result, Hitler ordered Yugoslavia to be invaded. In the event, the subjugation of the Balkans would result in a four-week delay of Operation Barbarossa, the invasion of the USSR.

On 5 April 1941, IV./KG b z b V 1 was based at Krumovo, Bulgaria, as part of VIII. Fliegerkorps. The following day, German and Italian troops invaded Yugoslavia and Greece, with the invasion being codenamed Operation Marita. The campaign lasted eleven days before the Yugoslav government surrendered. The Germans now turned their full attention to Greece. On 6 April, German forces based in Bulgaria had already attacked the north-eastern parts of the country.

By 14 April, three further *Transportgruppen*, KGr z b V 1, 2, and 3 respectively had arrived in Bulgaria, with German troops being airlifted into the Greek Korzani Plain. The Isthmus of Corinth was attacked on 26 April. Taking off from Plovdiv, the Ju 52/3ms made an intermediary landing at Larissa. Literally hugging the ground, the Ju 52/3ms flew three-abreast, with the exhaust flames of the BMW 132 engines guiding the aircraft next in line. Arriving undetected over the British positions (including large contingents of Greek, Australian, and New Zealand troops) immediately after a heavy Luftwaffe raid, the Ju 52/3ms disgorged its load of *Fallschirmjäger* from an altitude of 120 metres. A dozen DFS 230 assault gliders, towed by Ju 52/3ms of I./Luftlandegeschwader 1 landed shortly afterwards. By 1 May, more than 50,000 Commonwealth troops had been evacuated to Crete. Although an invasion of Crete was not part of the original Operation Marita, the Commander of XI. Fliegerkorps, *Generalleutnant* Kurt Student, proposed to take the island by air. The operation, subsequently approved by Hitler on 21 April, involved dropping thousands of *Fallschirmjäger* over strategic points.

With sufficient numbers of Ju 52/3ms available, some 3,000 to 4,000 *Fallschirmjäger* could be brought to Crete. However, many aircraft were in need of extensive overhaul. As a result, on 1 May, the Ju 52/3ms were ordered to fly north to overhaul centres. However, not enough fuel was available for conducting an immediate invasion of Crete—it was estimated that some 900,000 litres would be needed for the first sortie alone—although sufficient quantities of fuel eventually arrived by sea. With most of the airfields being some distance from the nearest port, the fuel then had to be transported by truck.

Two weeks later, on 15 May, no less than 493 Ju 52/3ms had been overhauled. The next objective was Crete. The airfields around Athens were now brimming with Ju 52/3ms. There were several problems, though; the airfields were covered in liberal amounts of dust, creating havoc with the engines of the Ju 52/3ms, and each take-off deposited so much dust into the air that it took on average seventeen minutes before it was deemed safe for the next aircraft to take-off. As a preventative measure, water was sprayed onto the soft to reduce the amount of dust. With no airfield support gear in place, each Ju 52/3m had to be refuelled by means of hand pumping fuel from barrels.

On 17 May, one Ju 52/3m of KGr z b V 1 crashed on landing at Melos.

Iraqi Diversion

When Iraqi forces, led by Rashid Ali, attacked the RAF base at Habbaniya in May 1941, Germany despatched a number of aircraft, including Heinkel He 111Hs and Messerschmitt Bf 110s of the VIII. Fliegerkorps in support of the rebellion. The detachment was named *Fliegerführer Irak*. A small transport force consisting of twenty Ju 52s, led by *Hauptmann* Harry Rother, as well as a few Ju 90s. Prior to leaving for Iraq, the Luftwaffe aircraft had their standard markings replaced by Royal Iraqi Air Force insignia. The first of the He 111s arrived in Baghdad on 11 May 1941; however, when Luftwaffe aircraft entered the landing pattern, they were fired upon by Iraqi tribesmen. One casualty was the officer, Axel von Blomberg, who intended to serve as liaison between the German forces and Rashid Ali. The German presence in Iraq was discovered the following day, with the first air combat occurring on 13 May.

One Ju 52/3m of *Fliegerführer Irak* was spotted at Palmyra in Vichy-French Syria by Tomahawks of 250 Squadron, which subsequently claimed to have damaged it during a strafing attack. On 20 May, one Ju 52/3m was damaged in a strafing attack against Mezze airfield, near Damascus—the Ju 52/3m was destroyed by fire. Four days later, one Ju 52/3m was damaged by 201 Squadron Blenheims at Al-Nayrab, near Aleppo.

One interesting incident took place on 28 May, when four Ju 52/3ms were escorted by Vichy-French MS 406 fighters. The dorsal gunners of the Ju 52/3ms did not appreciate the presence of the Vichy-French fighters, firing continuously until they had reached Al-Nayrab. None of the MS 406s were hit, though. In the event, the Iraqi rebellion failed, with the Germans losing in excess of twenty aircraft, including several Ju 52/3ms. Had the rebellion succeeded, the Germans would have gained a foothold in the strategically important Middle East, securing vital oil supplies. Incidentally, it is interesting to note that the fuel supplies for the *Fliegerführer Irak* had to be airlifted to Iraq by the Ju 52/3ms, as no oil refinery was available.

Operation Barbarossa

Due to the *Transportgruppen* being heavily committed in North Africa and the Balkans, only four units, IV./KGr z b V 1, KGr b z V 50, KGr z b V 102, and KGr b z V 106, were immediately available for Operation Barbarossa. The invasion of the USSR created a number of problems not encountered in Western Europe. The distances involved were far greater than in the west, with the roads more often than not being in poor condition, and the railways had a larger width of track than those in Germany. As a result, the dependence of air transport for the supply of advancing

A Ju 52/3m g7e between sorties. Note the twin machine-gun armament. (*Author's Collection*)

troops was considerably greater. On 17 December 1941, four new *Transportgruppen*, KGr z b V 700, 800, 900, and 999, were formed for operations on the Eastern Front. With German forces being halted just outside Moscow by stiff Russian resistance and 'General Winter', the Soviet forces would eventually begin to push back the Nazi invaders.

In early January 1942, the Red Army attacked German forces at Sukhinichi. Hitler, who in December 1941 had assumed overall command of the German armed forces, ordered a resupply operation, using Ju 52/3ms of KGr z b V 172, to the beleaguered Germans. During the operation, six Ju 52/3ms were claimed as destroyed on the ground in a Soviet bomb raid on 25 January, with another one shot down.

Demyansk Airlift

As the air supply operation at Sukhinichi was ongoing, some 100,000 German troops had been encircled at Demyansk. During three months, 35,400 sorties were flown. About 220 Ju 52/3ms were available, with some aircraft being transferred from *C-Schule* and other units (several Fw 200s and Ju 90s also took part). Among the units participating were the newly formed KGr z b V 600, 700, 800, 900, and 999. Between 19 February and 18 May, an average of 150 transport aircraft were used on a daily basis. In total, 24,303 tons of supplies arrived within the Demyansk pocket, as well as 24 million litres of petrol—the daily average was 273 metric tons. More than 15,000 troops were airlifted into the Demyansk pocket, with 22,000 wounded being flown to safety. Simultaneously, German forces at Kholm were under intense pressure, requiring resupply by air. With the Luftwaffe enjoying air supremacy, few Ju 52/3ms were lost during the initial stages of the air supply operations at Demyansk and Kholm. However, between 22 February and 10 March, sixty Ju 52/3ms were claimed as destroyed. Out of these, thirty-nine losses can be verified in German records.[3]

The final Ju 52/3m losses in the Demyansk pocket occurred on 8 April, when four were claimed as shot down by a 580 IAP fighter pilot (three were actually lost). By May, the siege was largely lifted,

A Ju 52/3m of KGr z b V 600, BR+, seen during the Demyansk airlift. (*Author's Collection*)

Just prior to landing at Demyansk, *Unteroffizier* Vogel's Ju 52/3m, G6+DX of KGr z b V 105, was attacked by a Soviet fighter. Vogel and his crew managed to descend in one piece, but their aircraft collided with a stationary Ju 52/3m—both aircraft were repaired. (*Author's Collection*)

with most *Transportverbände* being transferred to other sectors. Nevertheless, three transport units would carry on operations until October 1942. Total Ju 52/3m losses during the February–May 1942 Demyansk air supply were 106 aircraft destroyed or severely damaged. In the event, the success of the Demyansk and Kholm air supply operations would have dire consequences for the Germans later on during the Stalingrad airlift. Göring boasted that the Luftwaffe would be able to resupply by air any German forces as it had done at Demyansk and Kholm.

13
Stalingrad

For many, the Ju 52/3m is forever linked with the battle for Stalingrad. Hundreds of Ju 52/3ms, as well as other transport aircraft, delivered supplies to the encircled German 6th Army. In the event, the 6th Army surrendered in the largest German defeat on the Eastern Front thus far in the war. The German defeat at Stalingrad and El Alamein were the major turning points in the war.

However, it is necessary to briefly describe the events leading up to the airlift. The first Luftwaffe raid against Stalingrad occurred on 23 August. During the following days, the city was hit repeatedly by the Luftwaffe. Nearly 1,000 persons were killed in the raids, with the water and electricity supply being knocked out, as well as the railway station. It was expected that Stalingrad would fall to the advancing German forces; however, having the city of Stalingrad (named after Josef Stalin) fall into the hands of the Nazi invaders would not only be a military defeat of large-scale proportions, but also a huge propaganda loss to the Communist dictator—Stalingrad and its defenders would never surrender. The battle for Stalingrad was bloody and ruthless, with German and Soviet troops engaged in bitter fighting among the ruins of the virtually destroyed city. Nevertheless, the Germans (including their Hungarian, Italian, and Romanian allies) failed to break the back of Soviet resistance. On 19 November, a Soviet counter-offensive began, which resulted in the German 6th Army, led by General von Paulus, becoming encircled. An airlift operation was quickly organised. Göring boasted that the Luftwaffe would have little trouble in resupplying the beleaguered German ground forces. Others, such as *Generaloberst* Wolfram Freiherr von Richthofen, commander of *Luftflotte* 4 and *Generalleutnant* Martin Fiebig, commander of the VIII. Fliegerkorps, were highly sceptical for the possibility of success for the Stalingrad airlift. With Göring on a shopping spree in Paris, the first Ju 52/3ms landed on the airfields inside the Stalingrad cauldron on 25 November in sunny and calm conditions. On this day, Göring had stated that only 500 tons of supplies would be necessary to support the 6th Army. However, it was subsequently calculated that 700 tons of supplies would be required.

About 320 Ju 52/3ms (including aircraft transferred from a variety of second-line units) were immediately available for the airlift, as were thirty He 111 transports. However, many aircraft were unserviceable. Nevertheless, with the Ju 52/3m being able to carry 2 tons, and the He 111 1.2 tons, there was initially some optimism within the German High Command that the Stalingrad airlift would succeed, as it had done at Demyansk.

The Luftwaffe was largely unprepared for the onslaught of the Russian winter. Here, Luftwaffe personnel shovel snow beside a Ju 52/3m preparing for take-off. A second Ju 52/3m, seen in the background, is coming in to land. (*Author's Collection*)

Luftflotte 4 Ju 52/3m units, 30 November 1942

Unit	Base	Available aircraft	Serviceable aircraft
KGr z b V 9	Konstantinovka	43	15
KGr z b V 50	Tatsinskaya	33	13
I./KG z b V 172	Stalino – Nord	29	11
KGr z b V 500	Tatsinskaya	45	35
KGr z b V 700	Taganrog	53	53
KGr z b V 900	Tatsinskaya	41	12
Total:		244	139

Being of rugged construction and easy to maintain, the Ju 52/3ms were able to operate under the most appalling conditions. One Luftwaffe pilot, Werner Utter, later recalled that it was possible to maintain altitude up to 2,500 metres on only two working engines with a cargo capacity of 85 per cent. On one engine, it was possible to fly as far as 80 km if starting from an altitude of 2,000 metres. During the Stalingrad airlift operation, one Ju 52/3m had remained on the ground for about three weeks, and about 1 metre of snow covered the wings. After removing the snow, the propellers were pulled round a few times and the engines started with no problems, with the crew finally being able to fly the Ju 52/3m.

In the event, the quota of 700 tons deemed necessary, or even Göring's lower promise of a daily average of 500 tons, was impossible to sustain. On 19 December 1942, 136 Luftwaffe transports, including fifty Ju 52/3ms, delivered 270 tons of supplies. The following day, 291 tons were airlifted into the pocket. Soviet resistance was limited, with only five aircraft, including one Ju 52/3m, suffering damage from anti-aircraft fire. Early on Christmas Eve, Soviet tanks broke through the German lines, attacking Tatsinskaya airfield. Within twenty minutes, panic-stricken German crews evacuated 124 Ju 52/3ms to Novocherkassk, Salsk, and other airfields. This effort, in poor weather, was later described as 'the most splendid effort carried out by the Luftwaffe!'[1] However, fifty aircraft, including twenty-two Ju 52/3ms, were destroyed or captured in the attack, with 300 tons of fuel, five ammunition stores, as well as airfield ground support equipment falling into Russian hands. Nevertheless, Ju 52/3ms operating out of Salsk managed to reach Stalingrad on 25 December. The next day, 78 tons were delivered for the loss of five aircraft.

In early January 1943, *Generalleutnant* Martin Fiebig (who was in charge of the Stalingrad airlift operation) reported that no less than sixty-two Ju 52/3ms had been either destroyed or badly damaged between 28 December 1942 and 4 January 1943 (official Luftwaffe records list only forty-nine Ju 52/3ms lost). Out of these, about half of the losses were weather related. The actual number of air-supply sorties flown was severely reduced. For instance, only forty-four sorties were flown on 8 January, with nine Ju 52/3ms lost.

Between 24 November 1942 and 2 February 1943, the air-supply operations cost the Luftwaffe 266 Ju 52/3ms, roughly 30 per cent of the number available. About 3,400 sorties were flown into the Stalingrad pocket between 25 November 1942 and 2 February 1943, delivering 6,591 metric tons of supplies. On average, this meant 94 metric tons daily, far lower than the results achieved during the earlier Demyansk resupply operations. In the event, the massive effort to supply the beleaguered 6th Army by air failed completely. Nevertheless, the Ju 52/3ms and other transport aircraft had brought out 24,760 wounded troops, as well as 5,150 'key personnel'.[2]

With Stalingrad recaptured, some 90,000 German soldiers surrendered, being marched into a very harsh captivity. On the Eastern Front, and in the war as a whole, the tide had turned. From now on, Germany would have to fight against superior Allied forces on all fronts. On the Eastern Front, Ju 52/3ms of I./TG 2, III./TG 2, II./TG 3, III./TG 3, and I./TG 4 flew in support of German forces in southern Ukraine, including the Crimean Peninsula. On Christmas Eve 1943, a large-scale Soviet offensive across the southern part of the front began. By late January 1944, several German units in the Korsun-Shevchenkivskyi and Kamianets-Podilskyi areas had been surrounded.

As a result, Luftwaffe air supply operations were initiated on 31 January. During the first sortie to Korsun-Shevchenkivskyi, no less than twelve Ju 52/3ms were claimed as shot down by Soviet fighters. The unusually mild winter caused the snow to melt, turning the airfields into a muddy chaos. With few fighter aircraft available, the Ju 52/3m formations were regularly escorted by as few as three Bf 109s or Fw 190s. In the event, the German forces at Korsun-Shevchenkivskyi managed to break the siege. However, nearly 19,000 troops had been killed (including General Stemmermann), with all of the heavy weaponry being lost. The entire air supply operation lasted seventeen days, with thirty-two Ju 52/3ms lost to all causes and another 113 suffering damage. In total, 2,000 tons of supplies were delivered, with 2,000 wounded troops being evacuated by air.

Other air supply operations on the Eastern Front included supporting the First Panzer Army in southern Ukraine in the spring of 1944. The Ju 52/3m units included were I./TG 1, IV./TG 1, I./TG 3, and I./TG 4. Following the loss of the airstrip at Kamianets-Podilskyi on 2 April, all supplies were air-dropped during night sorties. As the German troops had been ordered to live off the land, the only supplies air-dropped were ammunition, fuel, and spare parts for the Panzers. With no Soviet night fighters operating in the area, and with largely inefficient anti-aircraft artillery, only a few transports were lost to enemy action.

By early April 1944, it had become obvious that Crimea would have to be evacuated. Airlift operations began on 12 April. However, two days later, Hitler declared that Sevastopol would be defended until the end. As a result, troop reinforcements were to be ferried to Sevastopol. However, a strong Soviet offensive had been launched two days prior, resulting in the Luftwaffe *Transportverbände* working overtime with delivering supplies to the garrison at Sevastopol and bringing out sick and wounded. Twelve Luftwaffe *Transportgruppen* based in Romania, including seven with Ju 52/3ms (I./TG 1, III./TG 1, IV./TG 1, I/.TG 2, II./TG 2, III./TG 2, III./TG 3, I./TG 4, as well as one *Staffel* from I./TG 3), were involved in the air evacuation of the Crimea. More than 21,000 German troops were evacuated by air before the Crimean Peninsula fell to advancing Soviet forces.

14
North African Defeat

With the British victory at El Alamein, and the landings in Vichy-French Algeria during Operation Torch in November 1942, the days of Axis presence in North Africa were numbered. Even though the German forces were able to push back the inexperienced US troops during heavy fighting in Tunisia, Allied resources in men and materiel were slowly but surely gaining the upper hand. As a result, the Luftwaffe made numerous supply flights to the German forces in North Africa, bringing back wounded soldiers. For instance, in January 1943, 15,415 troops, 4,728 tons of various supplies (including arms and ammunition), 149 tons of fuel, and 8.5 tons of food were flown to Tunisia.

Considering the overall Allied superiority in men and material, this was clearly inadequate. The Ju 52/3ms usually flew as low as 30 metres above the Mediterranean, in an attempt to avoid Allied radar and the attention of fighters. Messerschmitt Bf 109 and Bf 110 fighters were assigned as escorts, but there were never enough Luftwaffe fighters available to protect the vulnerable transports. The *Lufttransportführer Mittelmeer*, commanded by *Oberst* Rudolf Starke, held overall responsibility for the airlift operations; however, Starke was replaced in February 1943 by *Generalmajor* Ulrich Buchholz. Allied bomb raids against the North African airfields accounted for many Luftwaffe aircraft. In two days, 18 January and 22 January, no less than thirty-eight Ju 52/3ms were destroyed at El Aouina by RAF fighters, with an additional thirteen being damaged. A singular Luftwaffe Ju 52/3m success occurred on 12–13 February, when the dorsal gunner of a KGr b z V Frankfurt aircraft managed to shoot down a RAF night fighter. However, such incidents remained rare. On 4 February 1943, the British 8th Army entered Tunisia. Under pressure from British and US forces, airlift operations in support of the German Afrika Korps were inadequate. An analysis of the Luftwaffe airlift operations indicated the flight patterns of the Ju 52/3ms and other transport aircraft, as well as the importance of their role in the supply and evacuation of the Afrika Korps.

As a result, Operation Flax was initiated on 5 April, which was a dedicated effort by the Allies to effectively reduce the Luftwaffe airlift into Tunisia. At 8 a.m. on 5 April 1943, a large formation of Luftwaffe aircraft were spotted over Cape Bon.[1] The formation was estimated to consist of between fifty to seventy Ju 52/3ms, escorted by twenty Bf 109s, four Fw 190s, six Ju 87s, and one Fw 189. USAAF fighters of the 1st FG and 82nd FG attacked, claiming eighteen Ju 52/3ms, along with five Bf 109s, five Ju 87s, one Bf 110, one Me 210, and the Fw 189. In the heat of battle, the number of Luftwaffe aircraft involved was exaggerated, with only some thirty-one Ju 52/3ms, ten Bf 109s, six or seven Bf 110s, four Ju 87s, and one Fw 190 being involved. Out of these, the Luftwaffe lost twelve, or possibly fourteen, Ju 52/3ms, as well as two Bf 109s. Far worse was to come for the Luftwaffe *Transportverbände*.

Eleven Ju 52/3ms at a Sicilian airfield. An Italian SM.79 Sparviero can be seen in the lower right corner. (*Author's Collection*)

One of many Ju 52/3ms destroyed during the final stages of the North African campaign. (*Author's Collection*)

On 18 April, what has become known as the Palm Sunday Massacre took place. USAAF Curtiss P-40s intercepted a large formation of Luftwaffe aircraft, consisting of sixty-eight Ju 52/3ms escorted by Bf 109s and Italian Macchi C.202s. No less than twenty-four Ju 52/3ms were claimed as shot down, with an additional thirty-five making emergency landings or crashing—another nine were claimed as damaged. Luftwaffe records show thirty-three Ju 52/3ms lost on this occasion. Operation Flax lasted nearly three weeks, during which 123 Ju 52/3ms were lost. In the event, Göring ordered the transport units to operate at night only, thus reducing the risk of Allied fighters locating them. As a result, night operations reduced the number of sorties. It was clear to all concerned that the days of German presence in North Africa were numbered. In early May, the Ju 52/3m crews of the *Transportverbände* flew numerous sorties in a desperate effort to evacuate as many personnel as possible; however, on 12 May, the last German and Italian units surrendered.

The exact number of Ju 52/3ms lost may never be known with certainty. It is believed that between 8 November 1942 and 12 May 1943, in excess of 350 Luftwaffe transport aircraft were lost. The majority of these were Ju 52/3ms. The loss of so many transport aircraft and their crews were crippling to the *Transportverbände*, and was seriously detrimental to the overall German war effort.

As a consequence of the huge losses of trained crews and aircraft, the Luftwaffe *Transportverbände* units were reorganised in May 1943 as *Transportgeschwadern* and *Transportgruppen*, and the nominal role of auxiliary bombers was finally dispensed with—the strength of a *Transportgeschwader* was 217 aircraft. Each of the four *Gruppen* had four *Staffel*, each with twelve aircraft, with the *Geschwader Stab* having five and the *Gruppen Stab* the remaining five aircraft. In order to ease maintenance, each *Transportgruppe* had a *Flugbetriebskompanie* (literally maintenance unit) assigned (please refer to Appendix II and III for a list of the *Transportgruppen*.)

Many German evacuees had landed at the small island of Pantelleria, with the majority slipping away before the island fell to the Allies on 11 June—Sicily was next. With only some eighty Ju 52/3ms available in Italy, evacuation by air proved difficult. By 19 August, Sicily had been lost, with the Allies preparing to invade the mainland. Air attacks against Luftwaffe airfields saw many of the valuable Ju 52/3ms destroyed or damaged. For instance, III./TG4 lost eleven Ju 52/3ms in a bomb raid against Pratica di Mare on 17 September 1943.

Following the Italian surrender on 3 September 1943, the race was on to secure strategically important islands in the eastern Mediterranean occupied by Italian troops. The *Luftwaffenkommando* Südost, based in the Balkans, had already begun to prepare for a possible Allied invasion of Greece. The airfields at Salonica, Larissa, Janina, and Corfu were seen as places from where the Germans would initiate a counter-offensive in case of an Allied invasion. In the end, British forces managed to occupy Kos and Leros in the Dodecanese Islands. However, already on 3 and 4 October, Kos was recaptured by *Fallschirmjäger* units, dropped from Ju 52/3ms of 6./TG 4. Ju 52/3m strength was soon augmented by aircraft diverted from Italy. On 12 and 13 November 1943, 500 German paratroopers descended on Leros, dropped from about eighty Ju 52/3ms, and heavy fighting ensued with British forces retreating four days later.

A little-known fact is that during the autumn of 1943, three Spitfire Mk Vs (W3750, EP731, and EP754) were fitted with floats by Folland and shipped to Egypt. The intention was to base the Spitfires in the Greek Dodecanese Islands where they would be used to intercept Luftwaffe Ju 52/3ms that were supplying island garrisons. However, when German forces managed to evict British forces from the islands of Kos and Leros in November 1943, these plans were not proceeded with.

By the spring of 1944, large areas of Yugoslavia were under control by Yugoslav Partisans, led by Josip Broz (better known as Tito). A large-scale anti-partisan operation, code-named Operation *Rösselsprung* (Knight's Move) was set in motion by the Germans. It was hoped to capture Tito at his headquarters near Drvar. Large numbers of assault gliders were used, while *Fallschirmjäger* were to be air-dropped by Ju 52/3ms of II./TG and, possibly, II./TG 2. Launched on 25 May, Operation *Rösselsprung* was a failure. Although Tito's headquarters was captured, only his Jeep and uniform were found.

On 24 August 1944, the evacuation of Crete and the Aegean islands was initiated. This later developed into a military retreat from Greece, Albania, and parts of Yugoslavia, and many Ju 52/3ms were destroyed in RAF bomb raids against Greek airfields. On 14 October 1944, British forces entered Athens, however, many of the German garrisons on the various islands in the Aegean Sea did not surrender until May 1945.

15

On the Retreat

On the Western Front it had become clear that an Allied invasion of France would soon take place. Allied bomb raids against Luftwaffe airfields and defence installations were a daily occurrence. One such raid against Vélizy-Villacoublay on 5 February 1944 saw eleven Amiot-built Ju 52/3ms destroyed or damaged. On 31 May 1944, the Luftwaffe possessed 555 serviceable transport aircraft, the majority of which were Ju 52/3ms, but also included Heinkel He 111Hs, Messerschmitt Me 323s, Savoia-Marchetti SM 82s, and Leo 451s.

Luftwaffe Ju 52/3m Units, 31 May 1944

Unit	Total number of aircraft	Number serviceable aircraft	Command
II./TG 2	12	9	*Luftflotte* Reich
TGr 20	32	28	*Luftflotte* 5
Seetransportstaffel 2	10	9	*Luftflotte* 5
I./TG 3	48	44	*Luftflotte* 6
IV./TG 1	45	41	Luftwaffe *Kommando Südost*
II./TG 4	39	36	Luftwaffe *Kommando Südost*
Seetransportstaffel 1	12	11	Luftwaffe *Kommando Südost*
Seetransportstaffel 3	8	7	Luftwaffe *Kommando Südost*
I./TG 1	45	39	XIV. Fliegerkorps
II./TG 3	43	38	XIV. Fliegerkorps
I./TG 4	47	47	XIV. Fliegerkorps
TGr Bronkow	41	38	XIV. Fliegerkorps
Total:	382	347	

Losses to roaming Allied fighters grew following the Allied invasion of France on 6 June 1944. On 7 June, seven Ju 52/3ms were destroyed at Le Mans in a strafing attack, and a Ju 52/3m (W Nr 640750) of 7./ZG 1 was shot down by an Allied fighter off Cazaux the same day—all fifteen aboard were missing.

The Allied invasion of southern France, Operation Dragoon, began in August 1944. Operating from the US Navy aircraft carrier USS *Tulagi*, flying a Grumman F6F Hellcat, Ensign Edward Olszewski of VOF-1 claimed a pair of Ju 52/3ms destroyed near Orange on 21 August.

Soviet Il-2 Shturmovik aircraft attacking the Luftwaffe airfield at Krasnogorsk. One Ju 52/3m is already burning fiercely. Note the driverless horse carriage in the foreground. (*Author's Collection*)

On the Eastern Front, Soviet forces were gradually pushing back the Germans. In late August 1944, Romania changed sides, declaring war on Germany on the 25th. An attempt to transport reinforcements to Romania was made, with one Ju 52/3m being shot down when trying to land at Otopeni, and a number of Luftwaffe Ju 52/3ms were captured by Romanian forces (as related elsewhere). USAAF fighters and bombers based in Italy continued to attack targets in Hungary and southern Germany, and on 9 and 10 October 1944, the USAAF's 15th AF attacked Börgönd airfield in Hungary. Large numbers of Luftwaffe aircraft were based there, including Ju 52/3ms, of which many were destroyed or damaged. On 27 December, the Hungarian capital Budapest was encircled by Soviet forces; as a result, Luftwaffe and RHAF Ju 52/3ms based at Pápa and Szombathely were used to deliver supplies to the German garrison (70,000 strong), which also included many Hungarian troops. In total, 80 tons of supplies, including ammunition, food, petrol, and medical supplies, were deemed necessary, but the plight and despair of the civilian population was not taken into account. When the last airstrips were overrun on 9 January, the Ju 52/3ms continued to deliver supplies by parachute, most of which, however, ended up on the Soviet side—not all of the supplies were appreciated by the defenders. On 29 January, sweets and cigarettes were dropped to the encircled garrison.[1]

Exactly one month later, on 9 February, the final air-drops took place. With Budapest under Soviet control, several thousand German and Hungarian troops managed to break through the front line. Some of the troops were eventually spotted by a Luftwaffe reconnaissance aircraft, with a few supplies being delivered by air on 14–15 February. A total of thirty-six Ju 52/3ms had been lost during the siege of Budapest.

October 1944 was a crucial month for German forces in Finland and Norway. Finnish forces became involved in driving the remaining German forces out of the country after the country was forced to accept a truce with the USSR. Several Luftwaffe Ju 52/3ms were lost during the retreat, including three shot down by Soviet Airacobras on 23 October. With Soviet forces pursuing the retreating German units into northern Norway, key Luftwaffe and Wehrmacht personnel were evacuated from Kirkenes on 25 October. One Ju 52/3m floatplane of TGr 20 (7U+IK) was flown to Scotland; after being delayed by poor weather at Trondheim-Hommelvik on 25 October, the crew decided to head for Iceland instead of Tromsø. They did not reach Iceland, alighting 48 km off the coast of Aberdeen due to lack of fuel. The crew were picked up by a trawler and brought ashore, and the ultimate fate of the Ju 52/3m is unknown.[2]

16

Final Defeat

Following the conclusion of the Crimean air evacuation, several Ju 52/3m units were transferred to Germany, including I./TG 1, II./TG 3, I./TG 4, and III./TG 4. In addition, Ju 52/3ms requisitioned from Lufthansa were organised into the *Luftverkehrsgruppe* (Mot.). Based at Calau, the *Luftverkehrsgruppe* were to operate on communications duties within the rapidly shrinking Reich. Following commencement of Operation Overlord on 6 June 1944, the *Luftverkehrsgruppe* was the only transport unit remaining in France. Due to the destruction of nearly all railway lines, the importance of supply by air became even more important. Transferring ground crew and equipment of the Luftwaffe's fighter and fighter-bomber units, as well as delivering fuel, ammunition, and bombs, was the most important tasks of the *Transportverbände*.

The Ardennes counter-offensive included air-dropping 800 *Fallschirmjäger* behind the US Army lines. Code-named *Stösser* (Sparrow Hawk), the operation was initiated on 17 December 1944. The Ju 52/3m crews, as indeed many of the *Fallschirmjäger*, were inexperienced, having little or no combat experience. A total of sixty-seven Ju 52/3ms were used during the operation.

One Ju 52/3m crashed on take-off, with the maximum take-off weight having been exceeded by a handsome margin. Two more were shot down by night fighters, with eight more being lost to accurate anti-aircraft fire. Poor weather, including strong winds, meant that the *Fallschirmjäger* were dispersed over a large area. The *Fallschirmjäger* that reached their respective target areas tried their best to disrupt lines of communication and destroy various vehicles, however, with no radios available, no contact could be established with other German units. On 19 December, the order to retreat to German lines was issued. *Stösser* was the last time that the *Fallschirmjäger* were air-dropped into action. Following the failure of the Ardennes counter-offensive, only a limited number of air-supply sorties took place on the Western Front. In total, forty-three Ju 52/3m sorties were made in the Ruhr area, with eleven aircraft being lost. One notable aspect of the final air-supply operations is that by April 1945 the airfields at Ludwigslust and Werder (Havel) were the staging points for air-supply sorties on both the Western and Eastern Fronts.

On 10 February 1945, the Luftwaffe air transport units were yet again reorganised. The *Lufttransportchef der Wehrmacht beim OKL* was established, which meant that the units of the *Transportverbände* finally came under the command of a single staff. As described previously, each of the *Luftflotten* had had *Transportverbände* attached to them, thus making it more difficult than necessary to organise and coordinate any large-scale airlift operation. Led by *Generalmajor* Friedrich-Wilhelm Morzik, the primary task of the *Lufttransportchef der Wehrmacht beim OKL* was to coordinate and assign the remaining *Transportverbände* as required. The situation for the Luftwaffe, and Germany as a whole, was desperate.

One of three Ju 52/3ms claimed shot down by a 9th AF fighter near St. Vith, Belgium, during the Ardennes offensive. (*Author's Collection*)

On the Eastern Front, a number of strongly defended cities (including Posen, Schneidemühl (Piła), Küstrin (Kostrzyn nad Odrą), Arnswalde (Choszczno), Glogau (Głogów), Breslau (Wrocław), and Gruzdžiai) were encircled by advancing Soviet troops. The garrisons remaining in these enclaves of German resistance each demanded air supply, which put an inordinate strain on the Ju 52/3m units. The supply sorties were flown by night, in order to avoid being shot down by Soviet anti-aircraft guns. With Soviet forces encircling Breslau (present-day Wrocław in Poland), the Luftwaffe began to drop supplies to the beleaguered German forces. When the aircraft were able to land, wounded soldiers were carried, but the possibility to land at Breslau lasted only until 6–7 April 1945, and, on that night, three Ju 52/3ms brought out fifty-two wounded soldiers. Thereafter, all supplies were dropped by parachute, and the last defenders surrendered on 4 May. No less than 165 transport aircraft had been lost, most of which were Ju 52/3ms.

With the Allies having complete air supremacy, the *Transportverbände* had to perform supply sorties by night, but with roaming Allied night fighters even these were highly dangerous. The USAAF's 414th Night Fighter Squadron, operating Northrop P-61 Black Widows, claimed four Luftwaffe aircraft shot down between 7 and 10 April, including one Ju 52/3m.

The 422nd Night Fighter Squadron, also operating P-61 Black Widows, claimed three Ju 52/3ms shot down on 12 and 13 April 1945. The RAF was not to be outdone, with the Mosquitoes of 409 and 488 squadrons claiming a number of Ju 52/3ms shot down in April 1945.

A Ju 52/3m lies wrecked in a German airfield. In the foreground is the remains of a Heinkel He 111. (*Author's Collection*)

Luftwaffe Ju 52/3m Units, 25 April 1945

Unit	Base	Number of aircraft	Remarks
1./TG 1	Windau (Ventspils)	11	*Luftwaffenkommando* Kurland
III./TG 3	Wimsbach	37	*Luftwaffenkommando* 4
TGr 20	Oslo Airport, Fornebu	38	Luftwaffe General Norwegen
Seetransportstaffel 2	Hommelvik	7	Luftwaffe General Norwegen. Floatplanes
II./TG 2	Schrasslavitz	32	VIII. Fliegerkorps
III./TG 2	Neuenburg	24	VIII. Fliegerkorps
Stab & 2./TG 1	Tutow	14	*Luftflotte* Reich
3./TG 1	Pütnitz	15	*Luftflotte* Reich. Floatplanes
II./TG 3	Güstrow	34	*Luftflotte* Reich
Total:		212	

Only two Ju 52/3m units, I./TG 1 and II./TG 3, were available to deliver supplies to Berlin, and it is estimated that around 200 sorties were made, with the loss of up to thirty aircraft. However, there were no tangible results of this desperate and final task of the Luftwaffe's Ju 52/3m crews. With Hitler committing suicide on 30 April, the German capital surrendered on 2 May, although a few isolated pockets of German resistance continued the fight for a couple of days; however, Germany finally signed an unconditional surrender on 8 May. The risk of ending up in a Russian POW camp saw several Luftwaffe crews (and their family members) defecting to neutral Sweden. One such attempt that failed involved three drunken Germans who stole a Ju 52/3m in a desperate attempt to abscond to Sweden. They crashed soon after take-off from Værnes, Norway, coming down in the residence of Ola Saltnessand at Buvika, 25 km south of Trondheim.

The importance of the Ju 52/3m to the Luftwaffe and the German armed forces as a whole cannot be overstated. Without it, the arduous, but mundane job of moving men and materiel between bases would have been much more difficult. The Ju 52/3m bore the brunt of the airlift operations, from the Polish campaign of September 1939 to the last desperate attempts to deliver supplies to encircled German strongholds on the Eastern Front in the spring of 1945. Compared to the high-profile fighter and bomber pilots, the Ju 52/3m crews of the *Transportverbände* received relatively little in the way of attention in the public media. Nevertheless, forty-three *Transportverbände* pilots were awarded with the *Ritterkreuz* (Knight's Cross) between 1940 and 1945. Out of these, twelve were awarded during the Stalingrad airlift. Between September 1939 and February 1945, the Luftwaffe lost a staggering 5,926 transport aircraft, and the highest losses occurred during the third quarter of 1943, when no less than 637 transport aircraft were lost to all causes. Although the majority of these were Ju 52/3ms, other types such as the Focke-Wulf Fw 200, Messerschmitt Me 323, and Savoia-Marchetti SM.82s are included in this total.

Out of the thousands of Luftwaffe aircraft surrendered in May 1945, only a small proportion of these were Ju 52/3ms. Of these, about 150 were located in the British sector, including thirty-nine in Denmark, fifty-eight in Norway, at least one in Belgium, and ten at Schleswig Air Base (Jagel), Germany. The exact quantity of operational Ju 52/3ms surrendered on all fronts is difficult to ascertain, but probably did not reach beyond 250 aircraft.

Following the unconditional surrender of Germany, the hunt was on to find Messerschmitt Me 262s, Arado Ar 234s, and other high-performance aeroplanes of the Luftwaffe. Locating technical documentation as well as finding and interrogating aircraft constructors was a top priority. Although the Junkers Ju 52/3m was patently obsolete, this, as well as other transport aircraft, was deemed to be of interest as well. The Ju 52/3ms could be used in rebuilding regular airline infrastructure as well as equipping transport squadrons of national air arms until more capable aeroplanes would become available. One USAAF officer, Edward P. Warner, stated in a report that Allied losses of transport aircraft could be replaced through captured stocks of Luftwaffe aircraft. Warner wrote that if only the USAAF would have sufficient quantities of Douglas C-47s and C-54s then '…the Russians might take about 2,000 Ju-52 transports; the British about 200; the Chinese about 600; and the smaller United Nations (including such countries as Belgium, France, the Netherlands and Norway) about 600'.[1]

Warner's estimate of 3,400 surviving Ju 52/3ms was far in excess of the actual number available, with only slightly more than 2,800 being built between September 1939 and September 1944. Only between 150 and 250 serviceable Ju 52/3ms were surrendered in the British, French, and US zones of occupation. One might have thought that with the unconditional German surrender in May 1945, the Ju 52/3m would have largely vanished from the skies of Europe—this proved to be far from being the case. While the combat aircraft of the Luftwaffe had little value other than as scrap, various training, liaison, and transport aircraft, including Ju 52/3ms, were taken on charge by air arms and airlines of several countries. New operators of the venerable Ju 52/3m were found in Belgium, Czechoslovakia, France, Great Britain, the Netherlands, Norway, and Yugoslavia. Other countries, such as Portugal, Spain, and Sweden, also continued to operate the venerable Ju 52/3m.

Spares soon became a major issue in the continued Ju 52/3m operations. One example of this was that the mouldings for the tyres, made at Hannover, were reportedly burnt by the Allies. Additionally, the BMW 132 engines of existing Ju 52/3ms soon became difficult to maintain. (Incidentally, Brazilian Ju 52/3m operators ordered a batch of BMW 132 engines in 1946–47 from France, where it remained in production. It is unclear if any engines were actually delivered.)

17

Civilian Operators

The Junkers Ju 52/3m saw widespread service with a large number of airlines on all continents—during the 1930s, it was the most common airliner in Europe. Apart from the German Lufthansa, airlines in most European countries, including Austria, Belgium, Denmark, Estonia, Finland, Great Britain, Greece, Hungary, Italy, Poland, Romania, Spain, and Sweden, operated Ju 52/3ms. It was the mainstay of not only the German airline business, but also in many other countries as well. Due to the expansive policy of establishing airlines in South America and Asia, Ju 52/3ms carried passengers and cargo in Bolivia, Brazil, Ecuador, Peru, and China. South African Airways flew Ju 52/3ms, relinquishing them to the South African Air Force only following the outbreak of war.

During the war, Ju 52/3ms continued to operate with airlines in neutral Spain and Sweden. As late as April 1944, several Ju 52/3ms were supplied to the Turkish state airline. After the war, several Ju 52/3ms were taken into service in Britain, with nearly ninety operating with Air France, and several Ju 52/3ms also continued to fly in Argentina, Brazil, and Ecuador. With the availability of more capable and modern aircraft, particularly Douglas C-47s sold off cheaply by the USAAF, the heyday of the Ju 52/3m had passed. Nevertheless, the eminent bush-flying qualities of the Ju 52/3m meant that three were obtained by Sepik Airlines of Australia.

By the early 1960s, however, the Ju 52/3m had virtually disappeared. Lack of certain items, including tyres and lack of spares for the BMW 132 engines, as well as high operating costs, were also contributing factors. It is interesting to note that the final civilian operators of the venerable Ju 52/3m in Australia and Ecuador used it on low-frequency routes, in and out of primitive jungle airstrips, carrying both passengers and freight. This was similar to the use originally envisioned by Ernst Zindel and his design team in 1930.

However, since the early 1970s and the emergence of the Warbird movement, a number of Ju 52/3ms and CASA 352s have been operated for pleasure flights and air show appearances. Eight Ju 52/3ms/CASA 352s are currently airworthy.

Angola

During the autumn of 1939, it was announced that the *Divisão dos Transportes Aéreos* (Angola Air Services) had placed an order for one Ju 52/3m. To be powered by BMW 132-L engines, it was expected that the Ju 52/3m would arrive in September or October 1939. Presumably, this order was cancelled due to the outbreak of war.

Argentina

There were two major Ju 52/3m civilian operators in Argentina; the airline Aeroposta between 1937 and 1950, and the *Ministerio de Agricultura y Ganadería* for anti-locust operations between 1949 and the early 1960s.

Originally established as a state airline, Aeroposta was taken over in early 1937 by a syndicate led by Admiral Galindez, with the board of directors considering the purchase of Ju 52/3ms and Ju 86s. In the event, three Ju 52/3ms were delivered to Aeroposta in October 1937, being initially (although only briefly) registered as R-344 (W Nr 5824), R-345 (W Nr 5833), and R-346 (W Nr 5829). Aeroposta operated on a number of domestic routes, introducing a 25 per cent discount on the Buenos Aires to Rivadavia and Rio Grande routes. Although intended to last only until 1 September 1938, the discount was extended until the summer of 1939. The southernmost part of Argentina, Tierra del Fuego, was served by Aeroposta's Ju 52/3m, as were the Comodoro Rivadavia oil fields.

In 1938, the Ju 52/3ms were re-registered as LV-AAB, -BAB and, -CAB respectively, and, in 1942, in accordance with new regulations, re-registered as LV-AAG, -AAH, and -AAI. The Aeroposta Ju 52/3ms were based at Pacheco airfield near Buenos Aires, operating mainly in Patagonia.

Following an earthquake in Chile in January 1939, the governor of the Province of Santa Fe despatched an Aeroposta Ju 52/3m, LV-BAB, to Santiago on 30 January, carrying medical equipment and material for a field hospital.

One Aeroposta Ju 52/3m, LV-AAB, suffered repairable damage in an accident at Bahia Blanca on 9 February 1942. One additional aircraft was acquired from Syndicato Condor in late 1942, being registered as LV-AAJ (this was due to the fact that Argentina remained neutral during the Second World War, thus not being eligible for delivery of aircraft under Lend-Lease like virtually every other South American country). A fifth and final Ju 52/3m was bought from LAB in 1943, with Argentinian authorities giving permission for the ferry flight from Bolivia on 16 July 1943. However, for reasons unknown, but probably due to necessary refurbishment and lack of spares, it was not until 1945 that the Ju 52/3m was finally registered as LV-AAN. The final Ju 52/3ms of Aeroposta was withdrawn from use in 1950, being replaced by DC-3s bought second-hand in the USA, with three—LV-CAB, -AAJ, and -AAN—being transferred to the *Ministerio de Agricultura y Ganadería*.

LV-AAN of Aeroposta. (*Michael Magnusson*)

Anti-locust operations

With one FAA Ju 52/3m being tested as a crop duster in 1947, several aircraft were eventually transferred to the *Ministerio de Agricultura y Ganadería* in early 1949. The intention was to use the aircraft as anti-locust sprayers (incidentally, similar trials in South Africa in 1946–47 had determined that the Ju 52/3m was too big for such a task). The spraying flights involved flying at very low level, and several aircraft were ultimately lost in accidents. One such crash occurred on 12 March 1954, when LV-ZBE (W Nr 5656, a former Syndicato Condor aircraft) crashed soon after take-off from Catamarca. Nos 2 and 3 engines failed during the initial climb due to contaminated (water) fuel. The right wing struck a tree 3 km east of Catamarca, resulting in the total loss of the aircraft. Some of the Ju 52/3ms had their original BMW engines exchanged for Pratt & Whitney R-1340 units, for which spares were far easier to obtain. Nevertheless, the Ju 52/3ms continued in service until the early 1960s, with the final flight occurring probably on 14 August 1961.

Argentinian Civilian Ju 52/3ms

Registration	W Nr	Date registered	Name	P/i	Remarks
R-344	5824	08.10.1937	*Patagonia*	D-AALA	Re-registered LV-AAB 31.08.1942. Withdrawn from use (wfu) 1950, Cancelled 06.01.1954
R-345	5833	08.10.1937	*Pampa*	D-AYKA	Re-registered LV-AAH 31.08.1942. Suffered 10 per cent damage at Rio Grande 30.09.1946, repaired. Wfu 1950, cancelled 06.01.1954
R-346	5829	08.10.1937	*Tierra del Fuego*		Re-registered LV-AAI 31.08.1942. Re-registered LV-ZIF 18.04.1950. Suffered accident near Rauch 14.11.1956. Cancelled 10.08.1960
LV-AAJ	6800	26.12.1942	*Ibate*	D-AHGB, PP-CBP	Suffered accident at Lago Argentino 08.07.1946. Repaired. According to some sources, re-registered as LV-AXB, not confirmed. Re-registered LV-ZIG 18.04.1950. Re-registered LQ-ZIG 54. Wfu 56, Cancelled 10.08.1960

LV-AAN	4061	27.02.1945	*Quichua*	CB-21	Re-registered LV-ZIH 18.04.1950. Re-registered LQ-ZIH 54. Dbf at Tartagal 06.08.1953. Cancelled 04.06.1959
LV-YSH	4038	29.12.1950		D-ARUW, PP-AV, T-160	Re-registered LQ-YSH 54. Cancelled 10.01.1961
LV-YSI	6264	29.12.1950		164, T-164	Re-registered LQ-YSI 54. Cancelled 15.05.1963
LV-ZBD	4043	08.02.1949		D-3356, D-ABIS, PP-CAX, T-158	Re-registered LQ-ZBD 54. Preserved
LV-ZBE	5656	08.02.1949		D-AMYE, PP-CBL	W/o 3 km E Catamarca 12.03.1954
LV-ZBF	6403	08.02.1949		167, T-167	Re-registered LQ-ZBF? Ret to FAA 04.08.1954 as T-167
LV-ZBG	5053	08.02.1949		D-AQUQ, HC-SAC, PP-CBR, T-152	W/o El Chafar (La Rioja) 15.01.1953. Re-registered LQ-LBG 54! Cancelled 10.01.1961
LV-ZBH	4079	08.02.1949		D-APOR, PP-CBF, T-155	Cancelled 19.05.11954, ret to FAA as T-155
LV-ZBI	4075	08.02.1949		D-APEF, PP-CBG, T-156	W/o San Pedro de Colalao 10.03.1953
LV-ZBJ	6354	08.02.1949		D-AGCB, 168, T-168	Re-registered ZQ-LBJ 54. Cancelled 15.05.1963
LV-ZBL	6224	08.02.1949		166, T-166	Re-registered LQ-ZBL 54. Cancelled 10.01.1961
LV-ZBM	5120	08.02.1949		D-ADER, PP-CBE, T-157 or T-169?	Re-registered LQ-ZBM 54. W/o Esquel 14.01.1958. Cancelled 04.06.1959

Australia

In the mid-1930s, two Australian airlines stated their intention to purchase Ju 52/3ms. None were delivered, however. It was not until twenty years later that the then Australian mandate of New Guinea would be the scene of some of the last civilian commercial operations of the venerable Ju 52/3m.

In 1935, newly established Australian Transcontinental Airways Ltd announced plans to purchase six Ju 52/3ms, as well as two Junkers Ju 160s. It was intended to establish routes between Adelaide, Darwin, Melbourne, and Sydney. Although limited regular services were briefly maintained, using one Avro Ten and a GAL Monospar, no Junkers aircraft were acquired. Additionally, Southern Air Lines and Freighters Ltd intended to purchase four Ju 86s to be used for passenger services and three Ju 52/3ms for freight purposes. Both the Ju 86s and Ju 52/3ms were to be powered by Jumo 205 diesel engines—this did not happen. It wasn't until the mid-1950s that the Ju 52/3m made its first appearance in Australia.

Formed in December 1947 by former RAAF Spitfire ace Robert 'Bobby' Gibbes, Sepik Airlines operated in New Guinea. With many of the primitive airfields being unable to accommodate the larger DC-3, the Junkers Ju 52/3m was seen as the obvious answer. In 1955, three Ju 52/3ms were obtained in Sweden (this was not the first time that Gibbes had come across Ju 52/3ms. During the desert campaign in 1942–43, Gibbes had considered the Ju 52/3m a priority target as they ferried troops and supplies). All of them were parked at Bromma airport, with the asking price of £15,000 for each one of the transport aircraft. However, Gibbes later stated that he purchased all three of them for £1,500, including spares and support equipment. The work of refurbishing the Ju 52/3ms was led by Carl Nilsson, supplemented by former Luftwaffe technician Paul Raasch. Several BMW 132 engines were obtained from Finland, and a large batch of tyres from France.

The first of the Ju 52/3ms left Bromma on 8 October 1955, flown by Bobby Gibbes and Gustaf Adolf Jerdnell. It arrived at Goroka, New Guinea, on 18 November 1955. Originally built at Dessau in 1942, the Ju 52/3m had been operated by Lufthansa as D-AVIU, Aero O/Y as OH-LAM and in Sweden as SE-BUE. Immediately upon SE-BUE's arrival at Bankstown, Gibbes had the BMW engines replaced with Pratt & Whitney R-1340 Wasp engines, taken from CA Wirraway basic trainers. It was then registered as VH-BUU before entering service with Sepik Airlines. Gibbes considered the Ju 52/3m an excellent aeroplane:

> Operating the Junkers was like flying a Tiger Moth since it allowed you to do fighter-type approaches and landings and quite short, steep circuits and go straight in, whereas a DC-3 stood off some miles away and came in fairly low and slow.[1]

The last two Ju 52/3ms left Sweden on 13 January 1957. With Australia having supported Britain, France, and Israel during the recent Suez War, some problems were encountered in flying over Lebanon and Syria. Other problems during the epic flight were due to practical trials regarding the fuel consumption of the old Junkers. When flying over Iraq, the pilot of VH-BUV, 'Black-Jack' Walker, decided to find out exactly how long the Ju 52/3m would fly on the main fuel tank. Suddenly, one engine stopped, soon followed by the others. The crew aboard the second Ju 52/3m could only watch in horror as the Junkers slowly reached ground level. Fortunately, the centre engine started again; it turned out that, when the German-born flight engineer Paul Raasch had tried to change the fuel cock of the auxiliary tanks, the tap had broken off. Raasch finally managed to open the fuel cock when the Junkers was only about 200 metres from the inhospitable desert ground. Unsurprisingly, Walker refrained from such experimentation again.

The final two Ju 52/3ms were registered as VH-BUV and -BUW respectively. Being able to operate from short and rough jungle airstrips, the Ju 52s were flown and maintained under very primitive conditions. On 15 November 1958, Gibbes sold Sepik Airlines to Mandated Air Lines.

VH-BUV seen at Goroka, New Guinea. (*Brian Horseley via Dave McDonald*)

The first Australian Ju 52, VH-BUU did not last long, being lost in a crash near Baier River in October 1959. Although repairable, it was written-off as an insurance loss, and the wreck was partly destroyed later when intentionally set on fire by a film crew.

VH-BUW was damaged in a take-off accident at Wau, and the Junkers demolished the Quantas Terminal Building—the damage to the Ju 52 was confined to the left wing with the building being reduced to kindling. VH-BUW was subsequently dismantled and transported by road to Goroka where it was rebuilt. The first flight took place on 30 January 1959, and, in January 1959, VH-BUW was re-registered as VH-GSW, and was once again re-registered as VH-GSS on 25 August 1959. By September 1960, none of the Ju 52s remained airworthy, and the trio of Australian Ju 52/3ms were among the last of their kind to be operated by a commercial airline, with Ecuadorian airline Transportes Aéreos Orientales (TAO) becoming the sole Ju 52/3m operator. CH-GSS and -GSW were withdrawn from use and parked at Madang, before being cut up into manageable pieces and shipped to Australia for smelting. This job was performed by Brown's Scrap Metals of Adelaide—it was an ignominious end for such well-travelled and much-used aeroplanes.

Australian Civil Ju 52/3ms

W Nr	Variant	Registration	P/i	Delivered	Remarks
7256	12	VH-BUU	D-AVIU, OH-LAM, SE-BUE	08.10.55*	W/o 10.1959
7493	G8e/Z-5	VH-BUV	D-AXVA, OH-LAO, D-AEAC, OH-LAO, SE-BUD	13.01.57*	Re-registered as VH-GSS.
641375		VH-BUW	Lw, OY-DFU, SE-AYB	13.01.57*	Re-registered as VH-GSW.

* This date indicates when the aircraft left Sweden

Austria

Formed on 18 September 1923, Austrian Airlines (*Österreichische Luftverkehrs AG* (ÖLAG)) was a faithful operator of Junkers aeroplanes. Since its inception, ÖLAG had used Junkers F.13s, G 24s, as well as the G 31. By the early 1930s, Aspern Airport, Vienna, had developed into an important hub for the growing air traffic in Central Europe, being hosted by ten airlines. In 1934, two Ju 52/3ms were ordered by ÖLAG. The first of these, registered as OE-LAK, was delivered on 29 March 1935—the price was öS 253,269. Exactly two months after delivery, OE-LAK was slightly damaged when it struck a parked aeroplane on landing at Salzburg.

The second Ju 52/3m, OE-LAL, was delivered on 14 June 1935. With the delivery of the capable Ju 52/3ms, the number of passengers and freight carried increased rapidly. By the end of 1935, ÖLAG was the fourth largest airlines operating on the European continent after DLH, KLM, and Air France. Among the destinations were Salzburg, Graz, Klagenfurt am Wörthersee, Innsbruck, as well as several foreign cities, including Prague, Dresden, Berlin, Munich, Zürich, Venice, and Belgrade. On 16 March 1936, OE-LAL was lost during a flight to Klagenfurt am Wörthersee. Captain Mandl force-landed at an altitude of 2,000 metres near Wolfsberg in the Austrian Alps; fortunately, however, both crew and passengers survived, albeit most suffered injuries.

In 1936, an additional two Ju 52/3ms arrived, OE-LAM and -LAN, with a fifth one, OE-LAP, being delivered in 1937. The route network had expanded to include Laibach (present-day Ljubljana), Susak, Budapest, Sofia, and Thessaloniki. It was planned to expand the Thessaloniki leg to Athens, as well as inaugurating routes to Frankfurt, Brussels, and London. These plans were scrapped following the *Anschluss*, i.e. the incorporation of Austria into Germany in March 1938.

Interestingly, two Ju 52/3ms were added to ÖLAG's fleet after the Anschluss. The aircraft, registered as OE-LAR and -LAS, were delivered on 28 March 1938. When delivered, the Swastika was painted on the fins. In May 1938, all Austrian aeroplanes received German registrations, and,

An ÖLAG Ju 52/3m seen during the mid-1930s. (*Rob Mulder*)

on 17 December 1938, ÖLAG was formally incorporated into DLH, ceasing to exist on 1 January 1939. Most of the ÖLAG staff were transferred to the DLH Ostmark (later renamed Südost, south-east) branch. Of the seven ÖLAG Ju 52/3ms, six were taken over by DLH, and most of these subsequently ended up with the Luftwaffe.

ÖLAG Statistics 1935–1937

Year	Km flown	Passengers	Freight and luggage in kg	Mail in kg	Regularity in per cent
1935	553,803	12,610	247,288	30,269	99
1936	650,494	18,262	368,292	50,205	97
1937	795,048	19,501	441,251	81,351	97

Austrian Civil Ju 52/3ms

Registration	W Nr	Delivered	Remarks
OE-LAK	4076	29.03.1935	To DLH 1938, R D-AJAT and named *Raoul Stoisavljevic*
OE-LAL	5289	14.04.1935	W/o 16.03.1936
OE-LAM	4080	01.02.1936	To DLH 1938, R D-AGDA
OE-LAN	5590	27.06.1936	To D-AKEQ
OE-LAP	5727	30.04.1937	R D-ATEA
OE-LAR	5180	28.03.1938	R D-ALYL
OE-LAS	5933	28.03.1938	R D-AMFR

Belgium

Formed on 23 May 1923 as the Belgian Corporation for Air Navigation Services (*Société Aunonyme Belge d'Exploitation de la Navigation Aérienne* (SABENA)), the airline was Belgium's flag carrier until 2001. Before the outbreak of war, SABENA operated an extensive network of European and African routes.

In 1936, two Ju 52/3ms were ordered by SABENA. The first of these, registered as OO-AGU, entered service in late April, with the second, OO-AGV, arriving in Belgium the following month. The Ju 52/3ms were operated on European routes, including Amsterdam Airport Schiphol and Berlin Tempelhof. In all, eight Ju 52/3ms were acquired by SABENA, with the last one, OO-AUK, arriving in May 1939.

Two Ju 52/3ms were lost in accidents before the war. The first accident occurred on 16 November 1937, when OO-AUB hit an industrial chimney while on approach to Stene, Ostend. On 14 March 1939, OO-AUA was written-off in a crash at Haren-Evere. Already on 27 March 1938, OO-AGU was despatched to Belgian Congo. Upon arrival, a Congolese CofA, C-17, was issued.

When war broke out on 3 September 1939, the SABENA fleet included six Ju 52/3ms. Although limited scheduled flights to the Netherlands and Scandinavia continued, all operations were halted on 13 September. With no foreseeable use in Europe for the Ju 52/3ms, it was decided to fly them to the Belgian Congo. One obvious reason behind this decision was the risk of having the Ju 52/3ms misidentified as

OO-AGU of SABENA. (*Rob Mulder*)

Luftwaffe aircraft. With special permission from French authorities, the first three SABENA Ju 52/3ms were flown south in November 1939 with the remaining three following in January 1940.

Belgian Congo

With the arrival of half a dozen Ju 52/3ms in Belgian Congo, scheduled services began in earnest in early 1940. Two further Ju 52/3ms were obtained in late 1941 from BOAC. The first of these, the former G-AERU, was ferried to Belgian Congo on 30 September 1941. It was registered as OO-CAP, with the 'C' indicating the Belgian Congo register. The second BOAC Ju 52/3m, G-AERX, had to be put back into flying condition, which delayed delivery until 6 April 1942. It did not receive a civilian registration, being used instead as a source for spares.

Throughout the war, the SABENA Ju 52/3ms flew numerous support missions for the Allies on the African routes. Two were lost: OO-AUG, which crashed 80 miles north of Bangui (in present-day Central African Republic) on 1 January 1943, and OO-AUF, which was written-off on 3 April 1944 at Mongana. The final Ju 52/3m in SABENA service was OO-AGU. Having arrived in Congo already in 1938, -AGU had been damaged on 25 March 1944. The damage was apparently minor, as the Ju 52/3m was repaired. It was finally destroyed in an accident at Costermansville on 25 November 1947.

Post-war Use

Following the liberation of Belgium, the Belgian government sent a request to the Allied Commission involved in the disposing of former Luftwaffe aircraft, asking for five Ju 52/3ms. In the event, only four were delivered from Fuhlsbüttel, Hamburg, to Gütersloh on 24 April 1946. The Ju 52/3ms were to be operated by the *Ministerie van Verkeerswezen, Bestuur voor Luchtvaart/Ministère des Communications* (Ministry of Communications), and *Administration de l'Aéronautique* for calibrating aerial beacons around Belgium. Interestingly, one Ju 52/3m was registered as OO-SNA already on 18 April 1946. This particular Ju 52/3m has not been positively identified, although it may have been the one found at Sint-Denijs-Westrem, Ghent, where it had been abandoned by the retreating Germans.

Incidentally, one unserviceable Ju 352, serial AM 110, which had landed at Knokke during a ferry flight to Britain, is said to have been offered to the Belgian government, who refused to accept it. However, four Ju 52/3ms from RAF stocks were accepted. Three of the Ju 52/3ms were overhauled at Charleroi by

Avions Fairey, and were delivered to the Belgian Ministry of Communications. The Ju 52/3ms were used for different tasks, including ILS and SBA testing, photogrammetry, radar, and VHF check flights as well as for photo flights. The aircraft appear to have been used little, with OO-SNA logging 465 flight hours between 1947 and 1952. The last one, OO-SNA, was withdrawn from use in 1954.

In the early 1980s, a Ju 52/3m was acquired from Portugal. Although the initial plan was to restore it to flying condition, this fell through. The Ju 52/3m is currently on static display at the Brussels Air Museum.

Belgian Civilian Ju 52/3ms

W Nr	Registration	Date Registered	Cancelled	Remarks
5510	OO-AGU *	27.04.1936	13.02.48	W/o at Costermansville, Belgian Congo, 25.11.1947
5514	OO-AGV *	09.05.1936	25.11.1948	
5672	OO-AGW *	31.12.1936	25.11.1946	Cancelled 25.11.1946
5815	OO-AUA *	22.05.1937	22.03.1939	Cancelled 22.03.1939
5827	OO-AUB *	02.06.1937	01.12.1937	Cancelled 01.12.1937
5858	OO-AUF *	06.01.1938	13.02.1948	Cancelled 13.02.1948
6036	OO-AUG *	15.12.1938	14.03.1943	Cancelled 14.03.1943
6410	OO-AUK *	22.05.1939	15.11.1946	Cancelled 15.11.1946
640401	OO-SNA #	VN747	18.04.1946	Cancelled 17.04.1954
6527	OO-SNB #	VN751	13.08.1946	Cancelled 17.04.1954
130863	OO-SND #	VN750	29.03.1949	Cancelled 31.05.1951

* Indicates SABENA aircraft.

# Indicates Ministry of Communications aircraft.

A fourth former-RAF Ju 52/3m, VM930 (W Nr 3341), was also delivered in 1946; it was reserved as OO-SNC. However, it did not enter service, being reduced to spares instead.

Bolivia

Beginning in 1932, several Ju 52s were purchased by the Bolivian airline Lloyd Aero Boliviano (LAB). Founded in 1925 by Bolivian and German businessmen in La Paz and Cochabamba, their intention was to exploit the large quantities of rubber found in the Amazon basin to the cities of Santa Cruz and Trinidad. By the early 1930s, air transport had proved itself, with LAB operating a number of Junkers F.13 and W 34s. Having established a good relationship with Junkers, the first Ju 52 arrived in late 1932. This aircraft, a Ju 52/3m de (W Nr 4018), was shipped to Chile, then assembled and flown to La Paz in late November 1932. Named *Chorolque*, the Ju 52 soon proved its value.

Two additional Ju 52/3ms arrived at Cochabamba in the final days of January 1933. Both had been acquired following substantial contributions from Simon Patiño, an extremely wealthy tin ore magnate. Both aircraft were purchased second-hand from Junkers, and were shipped to Rio de Janeiro and, following assembly, flown to Puerto Suárez on the Bolivian-Brazilian border. The 1,600-km (1,000-mile) distance between Rio de Janeiro and Puerto Suárez was covered in an eight-hour non-stop flight. The Ju 52/3ms were W Nr 4008 and 4009 respectively, and had originally been built as single-engined Ju 52 ces, but converted to trimotor configuration on the assembly line. The Ju 52/3ms were named *Juan de Valle* and *Huanuni* respectively (with no official civilian aircraft register, the aircraft were identified by their respective names).

Powered by three BMW-built Hornet engines, the Ju 52/3ms could carry 2.5 tons of freight or seventeen passengers. With Bolivia being involved with Paraguay in what was known as the Gran Chaco War, the Ju 52s were requisitioned by the Bolivian air arm, carrying out numerous supply flights over the next three and a half years (see Chapter 18 for further details). In April 1935, a fourth Ju 52/3m arrived, which was civilian registered as CB-21. Two Ju 52s were subsequently lost in crashes, *Chorolque* in January 1936 and *Huanuni* in December 1937. One additional Ju 52/3m was registered as CB-22 in June 1937—this was W Nr 5623, named *Illampu*.

In 1939, another Ju 52/3m was ordered, but ultimately not delivered due to the outbreak of war in Europe. This Ju 52/3m was to have been named *Doctor Grether*. In Bolivia, the Ju 52/3m proved its worth as a rugged, dependable airliner, being able to operate under very primitive and difficult conditions. Incidentally, the wreck of *Juan de Valle* was discovered in the late 1980s. Plans to recover at least some parts of this historic aeroplane, the first Ju 52/3m built, were apparently suggested, but not carried through.

Bolivian Civilian Ju 52/3ms

W Nr	Delivered	Name	Registration	Remarks
4018	11.1932	*Chorolque*		W/o 01.1936
4008	01.1933	*Juan de Valle*		W/o 04.11.1940
4009	01.1933	*Huanuni*		W/o 12.1937
4061	04.1935	*Bolivar*	CB-21	Sold in 1943 to Aeroposta, R LV-AAN
5623	06.1937	*Illampu*	CB-22	Sold in 1944 to VASP, R PP-SPJ
?		(*Doctor Grether*)		Ordered in 1939, not delivered

Brazil

German interest in the South American airline business can be traced back to 1919. In this year, the Colombian-German Air Transport Society (*Sociedad Colombo Alemana de Transportes Aéreos* (SCADTA)) was formed by ex-patriate German businessmen in Colombia. Managed by Fritz Hammer, SCADTA opened up the vast interior of Colombia to commercial aviation, using Junkers F.13s, Dornier Wal flying boats, as well as other German-built aeroplanes. Although a success, SCADTA was sold to US interests in 1930. In order to promote the sale and use of German-built airliners abroad, the Condor Syndikat was formed on 5 May 1924. Led by Fritz Hammer, the Condor Syndikat successfully demonstrated a SCATDA Dornier Wal to Brazilian authorities in 1926. As a result, the first Brazilian scheduled airline, named Linha da Lagõa, was formed in February 1927. Operations were soon transferred to VARIG (see below). However, on 9 November 1927, the Syndicato Condor Ltd initiated operations. The new airline was, in fact, a subsidiary of Lufthansa.

Syndicato Condor soon developed a large route network, linking the coastal cities of Rio Grande in the south with Belém in the north. An inland route from São Paolo linked Rio Branco on the Bolivian-Peruvian border. The Uruguayan capital Montevideo, Buenos Aires, and Santiago in Chile were also served by Syndicato Condor. At Buenos Aires harbour, a 200-ton floating dock was used

PP-CBA of Syndicato Condor at Mendoza. (*Rob Mulder*)

for embarking and disembarking passengers and freight. The dock contained passenger waiting rooms, fuel, and water tanks, as well as a workshop. In 1936, offices at Quilmes airport were built to accommodate the Ju 52/3ms of Syndicato Condor.

No less than sixteen Ju 52/3ms were transferred to Syndicato Condor from Lufthansa, with the airline contributing greatly to the economic development of the vast Brazilian interior. By 1939, Syndicato Condor operated a route network, in collaboration with Lufthansa Südamerika, covering 15,000 kilometres. The longest route, 4,230 kilometres, was the Belem to Rio de Janeiro route. However, with the outbreak of war, German controlling interests in Syndicato Condor were reduced. Strategically, the USA could not afford to have a German-led airline in the Southern Hemisphere. Brazil was soon committed to the Allied cause, which resulted in the formation on 16 January 1943 of Serviços Aéreos Cruzeiro do Sul, which was wholly controlled by Brazilian interests.

VASP

Beginning in 1937, seven Ju 52/3ms, registered as PP-SPD, -SPE, -SPF, -SPG, -SPH, -SPI, and -SPJ, were purchased by *Viação Aérea São Paulo*, better known under the acronym VASP. As with Syndicato Condor and VARIG, VASP was controlled by German interests. Among the routes flown were São Paulo–Uberaba and Rio de Janeiro–São Paulo.

VARIG

Another Brazilian airline to operate the Ju 52/3m was *Viação Aérea RIo Grandense* (VARIG). Originally formed in 1927 at Port Alegre by ex-patriate German businessmen, VARIG received considerable assistance from Syndicato Condor. However, VARIG operations for the next fifteen years were limited to a number of local routes; although, by the early 1940s, American assistance saw a vast expansion of VARIG. A former SAA Ju 52/3m ge was acquired by VARIG (registered as PP-VAL), which crashed at Porto Alegre on 28 February 1942.

Most of the remaining Brazilian Ju 52/3ms were sold in 1946 to the Argentinian air arm. However, single examples continued in use until the late 1950s, when the final Ju 52/3ms were grounded. In recognition of the importance of the Ju 52/3m in developing regular air traffic in Brazil, an AAC.1 was obtained from Brazil for restoration and eventual display at the TAM Air Museum.

Brazilian Civilian Ju 52/3ms

Registration	W Nr	Name	P/i	Registered	Remarks
PP-CAT	4024	*Anhangá*		12.1933	W/o Rio de Janeiro 15.08.1938
PP-CAV	4038	*Caiçara*	D-ARUW	04.1934	Sold to Argentina as T-160. R LV-YSH
PP-CAX	4043	*Curupira*	D-3356, D-ABIS	03.1935	Sold to Argentina as T-158. R as LV-ZBD
PP-CAY	4042	*Marimbá*	D-3156, D-ASEN	04.1935	W/o Rio de Janeiro 13.01.1939
PP-CAZ	5261	*Maipo*	D-AGST	1937 *	Sold to Argentina as T-159
PP-CBA	5283	*Aconcagua*	D-AENF	29.09.1935	Re-registered D-AENF, OA-HHD, HC-SAD, 42-52883, TI-60
PP-CBB	4078	*Tupan*	D-AKIQ	12.1935	Sold to Argentina as T-153
PP-CBC	5453	*Guaracy*		11.1936	W/o Santos 22.05.1938
PP-CBD	5478	*Jacy*	D-AJAO	1937	Sold to Argentina as T-154
PP-CBE	5120	*Yarassú*	D-ADER	07.1938	Sold to Argentina as T-157 or T-169. R LV-ZBM
PP-CBF	4079	*Aracy*	D-APOR		Sold to Argentina as T-155, R LV-ZBH
PP-CBG	4075	*Pagé*	D-APEF		Sold to Argentina as T-156. R LV-ZBI
PP-CBH	5109	*Moré*	D-AMAQ		R HC-SAE, to FAP, R HC-SND
PP-CBL	5656	*Los Andes*	D-AMYE	09.1939	R LV-ZBE
PP-CBP	6800	*Ibaté*	D-AHGB		R LV-AAJ
PP-CBR	5053	*Uirapurú*	D-AQUQ	01.1939	R HC-SAC, re-registered D-AQUQ, re-registered PP-CBR, R LV-ZBG
PP-SPD	5459	*Cidade do São Paulo*		02.1937	W/o Rio de Janeiro 27.08.1943
PP-SPE	5465	*Cidade do Rio de Janeiro*		02.1937	Scrapped
PP-SPF	5689	*Cidade de Santos*		09.1937	Collided mid-air with DH 84 over Rio de Janeiro 08.11.1940
PP-SPG	6465	*Cidade de Goiania*			Scrapped São Paolo
PP-SPH	6510	*Cidade de Curitiba*			To AeroNorte re-registered PP-DZY. W/o Mogi das Cruzes São Paulo 1954
PP-SPI	6684	*Cidade de Cuiabe*	C-209		Re-registered PT-AUX W/o São Paulo 1957
PP-SPJ	5623	*Cidade de Campinas*	CB-22		
PP-VAL	4058	*Maua*	ZS-AFA	07.1938	Dbr at Porto Alegre 28.02.1942

Bulgaria

Following the end of the war, preparations for recommencement of regular civilian airline service began in December 1946 with the formation of the Directorate of Air Communications (*Direkcia na vazdushnite saobshtenia* (DVS)). Initially, three Ju 52/3ms were transferred from the Bulgarian Air Force, along with facilities at Burgas, Gorna Oryahovitsa, Plovdiv, and Sofia. In all, forty personnel, including pilots, radio operators, engineers, and mechanics, were seconded from the Air Force to DVS. On 29 June 1947, Ju 52/3m LZ-UNB was used to inaugurate the Sofia–Burgas route. Just over a month later, on 2 August, a second route, between Sofia and Varna was introduced. An additional three Ju 52/3ms arrived in February 1948. The first international route, to Budapest via Belgrade, was inaugurated on 13 February 1948, using Russian-supplied Li-2s. On 3 November 1948, TABSO was established, which was a joint-venture between Bulgaria and the USSR, and scheduled services began on 1 August 1949. The remaining Ju 52/3ms were returned to the Air Force in 1950.

With the DVS being responsible for the development of civilian aviation, including agricultural, airline services were handled by TABSO. At least one Ju 52/3m, registered as LZ-UNO (a Ju 52/3m g7e), was operated by TABSO, although it was withdrawn from use in the early 1950s.

Canada

During the 1920s, the need to transport people and cargo into the sparsely populated northern parts of Canada saw the emergence of 'bush' flying. One of the many airlines established during this time was Western Canada Airways Ltd. Operating out of Winnipeg, mining equipment, personnel, and various supplies were among the cargo carried. In November 1930, Western Canada Airways Ltd was incorporated into Canadian Airways. Western Canada Airways—and the subsequent Canadian Airways—were operating a number of all-metal Junkers F.13s and W 33/34s, which were excellent aircraft, but too small. There existed a need for a large cargo hauler, which would reduce the need to dismantle the bulky items that the Junkers F.13s and W 33/34s were unable to handle.

As a result, an active interest in the larger Junkers Ju 52 was shown. In the event, a second-hand Ju 52, originally registered as D-2137, was purchased, and the contract was signed on 3 July 1931. The following month, the head of maintenance for Canadian Airways, Mr T. W. Siers, travelled to Dessau to familiarise himself with the 685-hp BMW VII A inline engine. The sixth-production single-engined Ju 52 ce (W Nr 4006) was shipped across the Atlantic aboard the cargo ship *Beaverbrae*, arriving at Longueuil, near Montreal, Quebec, in November 1931. Following assembly by Fairchild Aviation Company, the Junkers was test flown by Alex Schneider on 28 November and was registered as CF-ARM. For obvious reasons, the Ju 52 was nicknamed *Flying Boxcar*. The cargo compartment was 19.5m^3 (690 cubic feet), with no less than nine hatches fitted. This considerably eased loading or discharging bulky cargo.

The Ju 52 was then demonstrated in various Canadian cities, carrying local politicians, civil servants, and industrialists aloft. It was also flown to Detroit, Michigan, where representatives of the Ford Motor Company inspected the German aircraft. However, despite the promise of rugged capability, it was severely shaken in 1932, when CF-ARM spent 297 days grounded due to insufficient lubrication of the engine. A representative of Canadian Airways referred to the Junkers as '…this white elephant'.

Plans to replace the engine were initiated, with the Pratt & Whitney Wasp being an option. In the event, the Rolls-Royce Buzzard was chosen. However, with the Buzzard out of production, a second-hand

The introduction of CF-ARM meant that larger and more bulky loads could be transported by air. (*Larry Milberry*)

engine was finally located in, of all places, Germany. The Buzzard was installed in January 1936, with the engine mount having to be modified. With the change of engine, the designation changed to Ju 52 cao. With the new engine, CF-ARM was proving its worth, so much in fact that a request was dispatched to Junkers about the availability of a second aircraft. Junkers replied that since the single-engined Ju 52 was out of production, instead recommending the purchase of the trimotor Ju 52/3m. Canadian Airways did briefly show interest in acquiring a Ju 52/3m powered by Wasp S3H1-G engines, but, mainly due to concerns for increased operating costs, nothing more came out of it.

During the summer months, CF-ARM was usually operated on floats, these being replaced by skis during the winter. On 17 December 1941, CF-ARM was sold to Canadian Pacific Railway. The Ju 52 was operated by Canadian Pacific Air Services Ltd. With the Buzzard engine about to expire, efforts to acquire another Buzzard began, but, with no Buzzards available in Canada, obtaining a Rolls-Royce Merlin was briefly considered; however, this was not possible. It is said that five Buzzards were located in Britain, but the ship carrying them to Canada was torpedoed.

In 1943, CF-ARM's Certificate of Airworthiness expired. By 1947, the Ju 52 was finally scrapped at Winnipeg. The fuselage became a plaything for children, later seeing use as a house trailer. Some small items, as well as the floats, still survive. A replica Ju 52 can be seen at the Western Canada Aviation Museum (for details on this aircraft, see Chapter 19).

Chile

The much-travelled D-AENF (W Nr 5283) was leased to Chilean national airline LAN Chile between 8 January and 14 March 1938. Soon after the return of the Ju 52/3m, LAN Chile received four Ju 86Z-1 airliners.

China

Since 1925, many Junkers aeroplanes had been supplied to China. The Chinese aviation ties with Germany were further strengthened on 21 February 1930, when the Chinese Ministry of Communications and

Lufthansa signed an agreement regarding the establishment of a joint Chinese-German airline. This airline was named Eurasia Aviation Corporation. Initially, a small number of Junkers F.13s and W 33s were delivered. In March 1934, the German ambassador to China contacted the RLM, requesting a Ju 52/3m being despatched to China. A DLH Ju 52/m, registered D-ABAN, was subsequently flown to China, with the flight lasting from 29 August to 6 September, with an airborne time of fifty-eight hours. The pilot in command was Carl August Freiherr von Gablentz.

The Ju 52/3m would demonstrate the capabilities of the new Junkers transport to Chinese authorities. One of the factors behind the arrival of the first Ju 52/3m in China was that, around the same time, a competing Chinese airline was introducing the American-designed Ford Trimotor into service. Showing off the more modern and capable Ju 52/3m was a logic marketing strategy. In the event, this particular Ju 52/3m did not enter service with Eurasia; instead, it was bought by the Nanjing Central government in April 1935 for use by Chiang Kai-shek as a VIP transport.[2]

Nevertheless, Eurasia did eventually operate no less than ten Ju 52/3ms, all of which had originally seen service with Lufthansa. They were registered as EU XV and XVII to XXV, with EU being short for Eurasia. The first of these was introduced into service in September 1935. At the Eurasia headquarters in Shanghai, a hangar containing space for three Ju 52/3ms as well as a staff room, workshop, radio room, and a 'pleasant reception hall' was erected.

Following the Japanese attack on 7 July 1937, Eurasia was forced to change its routes, also moving its headquarters, first to Xi'an in July and then to Kunming during the autumn. One Ju 52/3m, EU XVII, was slightly damaged in a Japanese bomb raid. The conflict with Japan saw a reduction of destinations, including Beijing, Nanking, and Shanghai. Nevertheless, on 31 December 1937, the Eurasia route network included Xian, Hanzhong, Chengdu, Hankou, Jiuduansha, Kunming, Canton, Hong Kong, and Hanoi in French Indochina. At Hanoi, it was possible to get on a connecting flight to Europe by means of Air France.

One remarkable episode occurred on 13 April 1939, when EU XIX was forced down by three Japanese bombers. The aircraft was on a scheduled flight from Hanoi to Kunming. The crew

Ju 52/3m EU XIX seen before taking off from the 325-metre makeshift runway. (*Author's Collection*)

managed to set down the aeroplane in one piece, and then walked for five days to safety, with the Ju 52/3m being left *in situ*. However, Chinese workers were recruited to literally hack a temporary 325-metre long runway from the side of the mountain where the Junkers had come down. On 15 October, the Ju 52/3m was flown off the runway, with the centre engine stopping soon after take-off.

The outbreak of war in Europe meant increased difficulty in securing spares and logistical support from Germany. Several of Eurasia's Ju 52/3ms had been lost in Japanese air attacks. However, one replacement Ju 52/3m arrived in August 1940. This did not last long, being shot down by Japanese fighters on 26 October. Some time before, on 18 April 1940, an air transport command using Eurasia aeroplanes was formed, although it was not taken over by the Chinese Nationalist Air Force (CNAF). The developing war situation saw a break in diplomatic ties with Germany on 2 July 1941. The Eurasia opening contract was withdrawn a month later, with all remaining Germans being ordered to leave China before 15 September.

On 1 June 1943, Eurasia was formally liquidated, with its assets being taken over by the Central Air Transport Corporation (CATCO). Only one Ju 52/3m remained, EU XX, in CATCO service, it was known as *Chung 5*. The Ju 52/3m lasted until 1945, when ample supplies of Douglas C-47s and Curtiss C-46s arrived.

Interestingly, one Eurasia Ju 52/3m was central to the plot in the novel *Band of Brothers*, written by the author (and pilot) Ernest K. Gann.

Chinese Civil Ju 52/3ms

W Nr	Registration	P/i	Delivered	Remarks
5329	EU XV	D-ANYK	09.1935	Destroyed 11.12.1941 at Kweilin in Japanese bomb raid
5104	EU XVII	D-AGES	10.1935	Destroyed 06.05.1939 at Hanzhong in Japanese strafing attack
5294	EU XVIII	D-AMAK	01.1936	W/o 01.08.1937 at Kunming
5472	EU XIX	D-AGEI	1936	W/o summer of 1945
5502	EU XX	D-ALUE	03.1937	Destroyed 30.12.1940 at Chengdu in Japanese bomb raid
5087	EU XXI	D-ASEV	05.1937	W/o 16.07.1938
4068	EU XXII	D-ABIZ	03.1938	Destroyed 08.12.1941 at Hong Kong in Japanese bomb raid
4074	EU XXIII	D-ASIS	09.1938	W/o 12.03.1939
4072	EU XXIV	D-AMIP	02.1939	Destroyed 08.12.1941 at Hong Kong in Japanese bomb raid
6014	EU XXV	D-AGFD	10.1940	S/d 26.10.1940 near Kunming

Colombia

One Ju 52/3m (W Nr 6684) was imported pre-war by the Colombia Petroleum Co. Registered as C-209, it was sold to Brazilian airline VASP on 14 May 1941.

The much-travelled *Aconcagua* seen in 1944–45 when used by the US Public Roads Administration. (*Dan Hagedorn Collection*)

Costa Rica

As related elsewhere, the solitary USAAF Ju 52/3m had proved to be virtually 'unassignable', with the aircraft eagerly being disposed of to the US Public Roads Administration in December 1943. With the Inter-American Highway under construction, the Ju 52/3m was deemed by the US Public Roads Administration to be an excellent tool in transporting survey crews and equipment in Costa Rica. It was duly registered as TI-60 in January 1944, remaining in service until the summer of 1945. Interestingly, the Brazilian airline Syndicato Condor then found out that one of their Ju 52/3ms (still officially registered in Brazil) was alive and well in Costa Rica. The US State Department advised the return of the Ju 52/3m, which, in the event, did not happen. Instead, the aircraft became the responsibility of the Foreign Liquidation Commission, who were responsible for disposing of US war equipment, including aircraft. The Ju 52/3m was sold in 1947 to the airline TACA. It remained in Costa Rica for a while, until transferred to TACA, Nicaragua.

Czechoslovakia

Immediately after the end of the war, the airline *Československé Letecká Spolecnost* (ČLS) briefly operated Ju 52/3ms. At least one flight, on 1 February 1946, was made to Bromma airport. Czech Airlines (*Československé Aerolinie* (ČSA)) restarted operations on 1 March 1946. Its first Ju 52/3m was transferred from the CVL (Czech Air Force). In all, ČSA operated three Ju 52/3ms (all refurbished by LETOV) and two AAC.1s between 1946 and 1948, using them on the Amsterdam, Belgrade, Berlin, Bratislava, Brussels, Karlovy Vary, Košice, London, Paris, Stockholm, Strasbourg, and Warsaw routes. One Ju 52/3m (registered OK-ZDN) crashed on landing at Prague on 5 March 1946. The Communist take-over in early 1948 resulted in Czechoslovakia being incorporated into the Soviet sphere of influence. On 18 July 1948, one ČSA Ju 52/3m (registered OK-ZDO) was flown to Austria (this flight was possibly connected with the defection of deputy Chief of Staff of the Czech Army, General Antonin Bohumil Hasal and his family to West Germany). The Ju 52/3ms and AAC.1s were eventually replaced in ČSA service by Douglas C-47s and Lisunov Li-2s.

A CSA AAC.1, registered as OK-PDC seen at Dübendorf, Switzerland, probably in 1946. (*Bernd Huckenbeck*)

Another Czech user was the Bata shoe company, which briefly operated a pair of AAC.1s, registered as OK-ZBB (c/n 111) and -ZBC (c/n 153) (one Ju 52/3m, registered as OK-ZBA, has also been reported). Incidentally, following the end of the war, the Czech government put forward a request to the RAF for Ju 52/3ms for use by the Bata shoe company. Three RAF Ju 52/3ms were delivered (RAF serials VM931, VN716, and VN745), but it seems likely that these were operated on behalf of the government instead.

Czech Civil Ju 52/3ms

Registration	W Nr	Remarks
OK-ODH		
OK-PCC *	36	Re-registered as OK-PDC
OK-TCB	6438	Re-registered as OK-TDI
OK-ZCA *	136	Re-registered as OK-ZDO. Defected to Austria 18.07.1948. Presumed returned
OK-ZDN		W/o Prague 05.03.1946
OK-ZBA		Unconfirmed
OK-ZBB *	111	Bata shoe company
OK-ZBC *	153	Bata shoe company

* Indicates AAC.1

Denmark

Following long discussions regarding the purchase of a new, more modern airliner, Danish Airlines (*Det Danske Luftfartselskab A/S* (DDL)) selected the Ju 52/3m over the Douglas DC-2. This was in no small part due to foreign currency exchange difficulties involved in a trans-Atlantic purchase. One Ju 52/3m (W Nr 5610) was bought new from Junkers, costing 485,000 kr. It was delivered on 27 July 1936, being put into service on the Stockholm–Copenhagen–Amsterdam route a few days later. Interestingly, in view of the limited amounts of US Dollars available, US-built Pratt & Whitney engines were specified for the new airliner. Formally registered as OY-DAL on 8 August 1936, the Ju 52/3m was named *Selandia.* It had originally been intended to name the Ju 52/3m in honour of the cabinet minister H. N. Andersen. However, as Andersen declined, the name *Selandia* was selected instead.

On 26 November 1936, OY-DAL sustained severe damage in a forced-landing. When flying over northern Germany, one propeller blade on the left engine suddenly broke away. The engine immediately began to vibrate badly, tearing itself off the engine mount. The engine remained hanging precariously on steel wires, with a fire developing around the carburettor. Captain Charles Bramsen had no recourse than to force-land north of Eutin. The Ju 52/3m was badly damaged, with one wing being ripped off from the fuselage, however, none of the crew or the thirteen passengers were injured. OY-DAL was sent to the Junkers factory in Leipzig, where it was rebuilt, returning to DDL service on 8 May 1937. One near-fatal incident occurred on 13 May 1938 when the starboard engine fell off over Falsterbo, Sweden. The crew continued to Copenhagen Airport, Kastrup, and managed to land safely.

While OY-DAL underwent repairs (returning to DDL on 23 June), a replacement Ju 52/3m, D-AJAN (W Nr 4026), was leased from Lufthansa. D-AJAN entered service with DDL on 28 May, and was returned to Lufthansa two months later, on 27 July 1938.

On the outbreak of war on 1 September 1939, OY-DAL was marked with large Danish flags beneath the wings and stabiliser, as well as the name *DANMARK* (DENMARK) in capital letters. The war also meant a sharp reduction in traffic for DDL. Following the German occupation of Denmark on 9 April 1940, the majority of the DDL fleet was grounded. The Ju 52/3m was flown to Lundtofte, where it was parked in the open. A barbed-wire fence was erected around the Junkers. Reduced air traffic was restarted on 24 June 1940.

On 18 December 1942, OY-DAL was lost. The Ju 52/3m crashed in poor weather shortly before landing at Aspern Airport. All aboard, including Captain Charles Bramsen, the wireless operator Sven-Aage Hansen, and sixteen passengers were killed on impact, and the aircraft was destroyed by fire—it was the first fatal DDL accident.

OY-DAL of DDL seen prior to take-off from Bromma airport, Sweden. (*Rob Mulder*)

Post-war

On 18 July 1945, an Aero O/Y Ju 52/3m, OH-LAM, was leased by DDL, being returned three weeks later. Three Ju 52/3ms were transferred to the Danish government from RAF stocks, with the transfer handled by the RAF's 8302 ADW—all aircraft were flown from Flensburg to Copenhagen Airport, Kastrup, by DDL crews on 1 April 1946 (W Nr 641375), 2 April (W Nr unknown), and on 16 April (possibly W Nr 641207). The Ju 52/3ms were all intended for DDL, although only one of the aircraft actually entered service, this being registered as OY-DFU on 28 October 1946. It was named *Uffe Viking*. The solitary DDL Ju 52/3m was used on the Copenhagen–Rønne route. Located on the small island of Bornholm, the airfield at Rønne could not accommodate Douglas DC-3s. One of the other Ju 52/3ms had the civilian registration OY-DFY reserved, but never entered service. Both the prospective OY-DFY and the third, unidentified Ju 52/3m was returned to the Danish government in December 1948, and subsequently allocated for firefighting practice at Kastrup.

OY-DFU was used until 4 February 1948, and stored at Kastrup awaiting a possible buyer. It was eventually sold to the Swedish company Aero Trafik on 23 August 1950, being ferried to Malmö Bulltofta Airport just over three weeks later, on 14 September. The Ju 52/3m was cancelled from the Danish civilian register on 30 September 1950, being registered as SE-AYB.

Danish Ju 52/3ms

Registration	Name	W Nr	Registered	Cancelled	Remarks
OY-DAL	*Selandia*	5610	08.08.1936	02.01.1943	W/o Vienna 18.12.1942
OY-DFU	*Uffe Viking*	641375	28.10.1946	30.09.1950	Sold to Sweden as SE-AYB
(OY-DFY)		641207 *			Noted derelict at Kastrup 18.09.1949
		?			Possibly Ju 52/3m MS variant

In addition, D-AJAN was leased by DDL from Lufthansa from 28 May to 27 July 1938

Ecuador

Formed on 24 July 1937, the *Sociedad Ecuatoriana de Transportes Aéreos* (SEDTA) was another South American airline established by German interests. Based at Quito, the formation of SEDTA was due to the influence of Fritz Hammer, a Syndicato Condor representative. However, Hammer was to die in the crash of a Junkers W 34 less than eight months later. Nevertheless, SEDTA continued to operate, although being largely dependent on monetary assistance from Lufthansa. The routes operated by SEDTA included Esmeraldas, Manta, Guayaquil, Salinas, as well as Distrito de Cuenca and Loja near the Peruvian border.

The first SEDTA Ju 52/3m, HC-SAB, did not last long, being written-off on 10 December 1938. Another Ju 52/3m, HC-SAC, was transferred to Syndicato Condor on 13 January 1939. In its place, HC-SAE was received from the Brazilian airline later that year, on 20 November. In May 1940, SEDTA put forward a proposal to open an air route to the Galapagos Islands; this made US authorities rather uneasy. As a result, the airline PANAGRA expanded its network of routes in order to operate in parallel with SEDTA. In spite of the growing tensions with the USA, two additional Ju 52/3ms were chartered by SEDTA from Syndicato Condor. One of these was the famous *Aconagua*; however, operations were severely hampered due to a fuel embargo. As a result of US pressure, the Ecuadorean government requisitioned the two remaining Ju 52/3ms, HC-SAD and -SAE on 5 September 1941. Both aircraft were passed to the *Fuerza Aérea Ecuatoriana* (Ecuadorean Air Force, FAE) with the former eventually being transferred to the USAAF in May 1942.

Post-war

Possibly due to the pre-war experiences of the Ju 52/3m (as well as availability), which had clearly demonstrated the ruggedness of the trimotor, the airline *Transportes Aéreos Orientales* S.A. (TAO) purchased a few aircraft in the 1950s. In fact, TAO would become the final commercial operator of the Junkers Ju 52/3m. Formed in 1948 and led by former Lufthansa airline Captain Cristof Drexel, TAO operated a combined passenger and cargo service across Ecuador. In 1952, TAO obtained the sole FAE Ju 52/3m, which was registered as HC-SND.

In the event, a second Ju 52/3m was added to TAO's fleet, being purchased from the Scandinavian airline SAS. In August 1957, the former LN-KAF arrived by boat. The Ju 52/3m was registered as HC-ABS and named *Amazonas*. Soon afterwards, on 2 April 1958, tragedy struck when HC-SND crashed shortly after take-off from Mariscal Sucre International Airport, Quito, after experiencing engine trouble. Out of the fourteen aboard, three were killed. Nevertheless, TAO continued to use HC-ABS until its eventual retirement in 1963.

The Ju 52/3m was not scrapped, instead being pushed into a corner of Quito-Mariscal Sucre airport and left to the elements. Here, it remained unused and unwanted for six years; however, this was not the end of the Junkers. It was purchased by American enthusiast Lester Weaver on 25 May 1970 and restored to airworthiness (see Chapter 18 for more details on this particular Ju 52/3m). Quito-Mariscal Sucre airport.

Ecuadorean Civil Ju 52/3ms

Registration	W Nr	Name	P/i	Registered	Remarks
HC-SAB	5915	*Ecuador*	D-APDF	1938	W/o Mt Chimborazo 10.12.1938
HC-SAC	5053	*Guayas*	D-AQUQ	02.1939	Sold to Syndicato Condor 14.08.1941 R PP-CBR
HC-SAD	5283	*Aconcagua*	D-AENF	11.1939	Re-registered D-AENF. Impounded by Ecuadorian government on 05.09.1941. To FAE. Transferred to USAAF 05.1942
HC-SAE	5109	*Azuay*	D-AMAQ, PP-CBH	06.1941	Impounded by Ecuadorian government on 05.09.1941. To FAE. Sold to TAO, R HC-SND. W/o 02.04.1958.

HC-ABS	5483/130714	*Amazonas*	D-AQUI, LN-DAH, Lw, D-AQUI, LN-KAF	08.1957	Retired in 1963. Sold to USA. Currently airworthy with Lufthansa *Traditionsflug* as D-CHLH

In addition, PP-CBG was chartered from Syndicato Condor on 10 February 1941. It was returned to Brazil shortly afterwards.

Estonia

As a result of a bill on civilian air transport passed by the Estonian Parliament in 1938, the airline AGO was formed on 6 February 1939. One of the founders was Jüri Ots, who in 1918 had established the Estonian Air Force. Both the Estonian State and private interests held shares in AGO, with the former having a share slightly more than 35 per cent. It was expected that AGO would initiate operations in August or September 1939. Two Ju 52/3ms were delivered to AGO, being registered as ES-AUL (W Nr 6613) and -AGO (W Nr 6633) respectively, and based at Lasnamäi Airport, Tallinn; however, by this time, war had broken out in Europe. AGO nevertheless initiated the route Tallinn–Helsinki–Stockholm (AGO route No. 1876) on 2 April 1940, which was flown three times weekly. The names of only two AGO pilots are known, Olt and Habel. The final flight on the Stockholm route took place on 14 June 1940, and two passengers were aboard the Ju 52/3m for the return flight to Tallinn. Following the Soviet occupation of Estonia, AGO ceased operations in July 1940. Both Ju 52/3ms were subsequently taken over in September 1940 by Aeroflot's Baltic Department. After

A poor-quality photo showing former LOT Lockheed 14 SP-BPN, the Estonian-designed PN-3 s/n 160, and one of the AGO Ju 52/3ms at Lasnamäe Airfield, Tallinn, following the Soviet take-over of Estonia in June 1940. (*Paul Branke*)

the German invasion, one of the Ju 52/3ms is said to have carried 600 rifles in aid of the defence of Leningrad, also transporting twenty-eight children to Gorki.

Estonian Civil Ju 52/3ms

Registration	W Nr	P/i	Date registered	Remarks
ES-AGO	6613	D-AXWA, ES-ALK	10.1939	To Aeroflot 09.1940. R CCCP-Л23 20.02.1941
ES-ALK	6613	D-AXWA	10.1939	Re-registered as ES-AGO
ES-AUL	6633	D-AXWB	10.1939	To Aeroflot 09.1940. R CCCP-Л22 20.02.1941

Finland

The Finnish airline Aero O/Y was a very early customer of the Ju 52/3m, placing an order for a single aeroplane on 7 January 1932. The original contract called for the aeroplane, W Nr 4014, to be delivered on 20 April, but the delivery was delayed due to financial difficulties encountered by Junkers. Additionally, the purchase of the Pratt & Whitney T2D-1 engines requested by Aero O/Y was met with some difficulty. However, on 24 June, the Ju 52/3m ce was handed over to Aero O/Y at Warnemünde, where the wheel undercarriage had been exchanged for floats. Registered as OH-ALK and named *Sampo*, the Ju 52/3m arrived at Helsinki some five days later, on 29 June. When fitted with floats, there was accommodation for fourteen passengers, but sixteen passengers could be carried with a conventional wheel undercarriage. The first service on the Helsinki–Turku–Stockholm route was flown on 1 July. By the end of the year, OH-ALK was leased to Swedish Airlines (ABA), not being returned to Aero O/Y until the summer of 1933. Another lease took place in October 1934, and the Ju 52/3m was returned in January 1935; with regular air traffic reduced during the winters, leasing the Ju 52/3m to ABA made good economic sense. The Ju 52/3m was touted as a 'modern, large aeroplane [that] will reduce bumping and airsickness to a minimum'.[3]

OH-ALK was damaged in a landing accident at Turku in December 1936, being repaired at Tampere. A second Ju 52/3m ge (powered by Hornet T2E engines) was ordered on 14 January 1936 and handed over to Aero O/Y at Dessau on 14 July 1936, and was registered as OH-ALL and named *Kaleva*. One remarkable incident occurred on 10 November 1937, when the centre engine fell away during a regular flight between Turku and Stockholm. Captain Helge Laitinen and the flight engineer Evald Hartikainen managed to keep the aeroplane straight and level. With the front windscreen being completely covered in oil, an *Ilmavoimat* (Finnish Air Force) pilot sitting in the cabin opened the dorsal hatch, shouting directions to Laitinen. Some of the lower fuselage was also torn off, providing the crew with a good view of the Baltic Sea below them. Fortunately, a safe landing was made at Turku, with some of the passengers enquiring as to the exact reason why they had turned back. Laitinen pointed to the nose of the Ju 52/3m, saying: 'Unfortunately, gentlemen, we had some engine trouble.' A reward was offered for information about where the missing engine and propeller had come down, but this produced no results. It was thought that the one of the propeller blades had cracked, resulting in the engine being torn from its mounting.[4]

The second Aero O/Y Ju 52/3m, OH-ALL. The location is most likely Lindarängen water airport, Stockholm. *(Swedish Aviation Historical Society)*

Following the outbreak of the Winter War on 30 November 1939, OH-ALK was pressed into service with *Ilmavoimat* (Finnish Air Force). Upon transfer on 1 December, OH-ALK had accumulated 6,468 hours, and on the same day the Ju 52/3m was re-registered as OH-LAK. It was returned to Aero O/Y on 21 February 1940, and was replaced by OH-ALL. However, the intention to re-register OH-ALL as OH-LAL did not take place, returning to Aero O/Y on 19 March 1940, with regular services resuming on 2 April. On 14 June 1940, OH-ALL was shot down over the Gulf of Finland near Ker lighthouse by a pair of Soviet Tupolev SB bombers. The two crew and seven passengers (one American, one Estonian, two French, two Germans, and one Swede) were all killed. Estonian fishermen reached the crash site, recovering bodies and some of the cargo, including diplomatic pouches. Soon afterwards, a Soviet submarine confiscated the diplomatic pouches, ignoring the bodies of the crew and passengers.

On 7 November 1941, OH-LAK crashed into the Baltic Sea off Turku, with two passengers losing their lives. The Ju 52/3m was recovered, and shipped to Königsberg for repairs, where it was fitted with BMW 132Z-3 engines, being returned to service on 15 October 1942. It was eventually written-off on 31 October 1945 in a non-fatal accident at Hyvinkää.

Three further Ju 52/3ms were delivered to Aero O/Y, the first of which was OH-LAM, arriving in December 1941. This particular Ju 52/3m-12 (a civilian production variant of the basic Ju 52/3m) had originally been intended for Lufthansa as D-AVIU, but was diverted to Aero O/Y. Named *Karjala*, OH-LAM was primarily used on the Helsinki–Kaunas service. When landing at Malmi Airport, Helsinki, on 6 July 1943, OH-LAM struck an *Ilmavoimat* Tupolev SB bomber. The rugged Ju 52/3m was repaired, with the Tupolev being struck off charge.

Due to the worsening war situation, as well as lack of fuel, most of the Aero O/Y routes were cancelled in the summer of 1944, and the last scheduled flight to Berlin took place on 3 August. In September 1944, the name *Karjala* was deleted (the Province of Karjala, in the eastern part of the country, was ceded to the USSR following the signing of the Finnish-Soviet peace treaty). On 18 July 1945, OH-LAM was leased to DDL, being returned on 13 August. OH-LAM was withdrawn from use in April 1949, remaining stored at Malmi Airport, Helsinki, until it was sold to Sweden on 8 November 1951.

The last two Ju 52/3ms, both Z-5 models built at Bernburg, were registered as OH-LAO and -LAP respectively, with the former entering service on 14 September 1942. It was leased to Lufthansa between 31 October 1942 and 3 July 1944, being registered as D-AEAC. The last flight in Aero O/Y service took place on 11–12 April 1949, with the Ju 52/3m being consigned to outside storage at Malmi Airport. As with OH-LAM, -LAO was sold to Sweden in 1951.

The final Aero O/Y Ju 52/3m, OH-LAP, was taken on charge on 26 August 1942. However, it was leased to Lufthansa on several occasions as D-ACDA. It seems likely that it was surrendered to US forces at Zickenbach, Austria, on 7 May 1945. Two days later, it was intentionally destroyed by machine-gun fire.

Finnish Civilian Ju 52/3ms

Registration	W Nr	P/i	Date registered	Name	Date cancelled	Remarks
OH-ALK	4014		08.07.1932	*Sampo*	01.12.1939	Re-registered OH-LAK. Leased to *Ilmavoimat* 01.12.1939–21.02.1940. W/o Hyvinkää, 31.10.1945
OH-ALL	5994		03.07.1936	*Kaleva*	14.06.1940	Leased to *Ilmavoimat* 21.02.1940–19.03.1940. S/d 14.06.1940 by Soviet SB-2 bombers
OH-LAM	7256	D-AVIU	19.02.1942	*Karjala*	30.04.1952	R SE-BUE
OH-LAO	7493	D-AXVA	12.09.1942	*Waasa*	18.01.1952	R SE-BUD
OH-LAP	7490		25.08.1942	*Petsamo*	1945	Leased to Lufthansa and R D-ACDA. Not returned

France

Following the liberation of France, the Amiot production line that had previously churned out Ju 52/3ms for the Luftwaffe was taken over for French use. The national airline, Air France, was the recipient of French-built AAC.1s already in December 1944. The first aircraft, registered as F-BAJA (carrying two construction numbers, 539 and No. 001) was taken on charge on 8 December. In the event, no less than eighty-seven AAC.1s were to serve with Air France. The AAC.1 was instrumental in re-establishing Air France on the European, North African, and Middle Eastern routes. However, due to its limited capacity and high operating cost, only thirty-five remained in use in early 1948. Many of the Air France AAC.1s had by this time been transferred to *Armée de l'Air*. The remaining Air France AAC.1s were based in Algiers, Brazzaville, Dakar, and Madagascar, with none

A close-up of AAC.1 F-BALE of Air France. Delivered to Air France in July 1945, the aeroplane was based in Madagascar; it was ultimately written-off on 10 April 1953. Note the half horse/half mermaid insignia on the nose, dubbed '*The Shrimp*' by Air France personnel. (*Collection Jean Delmas via Jean-Pierre Dubois*)

remaining in service on the European routes. In France's African colonies, the AAC.1s lasted for several more years, with the ones based in Madagascar remaining in use until the mid-1950s. Apart from carrying passengers, several AAC.1s were operated on the night-mail routes.

The AAC.1 could give outside observers a sense of ruggedness and, in some cases, having the appearance of being ready for the scrapyard. Nevertheless, in spite of primitive airport facilities, the AAC.1s would usually be able to deliver their passengers to their destinations safely, albeit slowly.

In his autobiography Camera in the Sky (Temple Press Ltd, 1958), the British aerial photographer Charles Sims describes an encounter with a French AAC.1 at Maison Blanche, Algeria:

> While we were on the ground, I strolled around the aerodrome with Harry Hensser, B.O.A.C.'s chief photographer and an old friend of mine. We found many relics of the North African campaign still there, including derelict British, American and German aircraft, and came across a very dilapidated and what we thought to be unserviceable Junkers Ju 52 Transport—a type of aircraft which to the Germans in the war was what the D.C.3 had been to us. This particular one had French roundels painted on it as well as civil registration letters.
>
> While we were wondering what its history had been, a very old motor coach rolled up and stopped alongside it. To our amazement about twenty people trooped out and climbed into the Junkers, the last of them a man in blue overalls, cap and choker, who turned out to be the pilot. He had also been the driver of the bus. After a few minutes while the people seated themselves and the 'driver' climbed up in front, the engines were started. Great clouds of blue smoke emerged, accompanied by loud pops and bangs, and

> from inside the fuselage somebody tried very hard to close the ill-fitting door. When all three engines had come to life and a steadier note had developed, the aircraft started to move forward towards the runway. We watched fascinated, amazed by the confidence of the twenty or so occupants of this dilapidated metal box with wings. On the runway and ready to take off, there was no nonsense about cockpit drill—the three engines were opened up, leaving a trail of blue-black smoke, and the aircraft just managed to stagger off the end of the runway. Slowly it disappeared from sight over the sea in the direction of Tunis, gaining a little height perhaps and, for as long as we could see it, somebody inside was still trying to close that door![5]

Although funny in a sense, the above story shows the harsh realities of regular passenger airline traffic in the immediate post-war years, as well as the fearless determination of both passengers and crew alike.

The Air France AAC.1s comprised: F-BAJA (001), F-BAJB (002), F-BAJC (003), F-BAJD (004), F-BAJE (005), F-BAJG (006), F-BAJH (007), F-BAJI (008), F-BAJJ (013), F-BAJK (014), F-BAJL (015), F-BAJM (016), F-BAJN (017), F-BAJO (018), F-BAJP (019), F-BAJS (020), F-BAJT (044), F-BAJU (036), F-BAJV (037), F-BAJX (038), F-BAKK (101), F-BAKL (102), F-BAKM (055), F-BAKN (056), F-BAKO (057), F-BAKP (058), F-BAKQ (059), F-BAKR (060), F-BAKS (071), F-BAKT (072), F-BAKU (073), F-BAKV (074), F-BAKX (075), F-BAKY (076), F-BAKZ (077), F-BALA (078), F-BALB (079), F-BALC (080), F-BALD (103), F-BALE (090), F-BALF (091), F-BALG (092), F-BALH (093), F-BALI (094), F-BALJ (095), F-BALK (096), F-BALL (097), F-BALM (098), F-BALN (099), F-BALO (100), F-BAMO (104), F-BAMP (105), F-BAMQ (106), F-BAMR (107), F-BAMS (108), F-BAMT (109), F-BAMU (110), F-BAMV (111), F-BAMX (112), F-BAMY (113), F-BAMZ (114), F-BANA (115), F-BANB (136), F-BANC (137), F-BAND (138), F-BANE (139), F-BANF (140), F-BANG (141), F-BANH (142), F-BANI (143), F-BANJ (144), F-BANK (145), F-BANL (146), F-BANM (147), F-BANN (148), F-BANO (149), F-BANP (150), F-BANQ (151), F-BANR (152), F-BANS (153), F-BBYF (226), F-BBYG (227), F-BBYH (228), F-BBYI (226), F-BBYJ (195), F-BCHJ (315), F-BCHP (321), and F-BDYK (148). A few aircraft, e.g. F-BANN (148), were transferred to another operator, and later returned, receiving a different registration, in this case F-BDYK (for more details on these aircraft, please refer to Appendix IV).

Non-scheduled Airline Operations

Apart from use by Air France, the AAC.1 was operated by a fairly large number of non-scheduled airlines. Most of these companies were short-lived, operating on routes in France, North Africa, and Indochina. Several aircraft were owned by a successive number of different companies. A brief rundown provides the following details:

Ste Aéro Cargo:	Based at Lyon, *Aéro Cargo* operated a single AAC.1—F-BCHH (323). It was written-off at Vichy on 14 July 1947.
Air Atlas:	This North African-based airline had six AAC.1s—F-BBZL (230), F-BBYT (240), F-BCHJ (315), F-BCHK (316), F-BCHL (317), and F-BCHQ (322).
Aigle Azur:	Had five AAC.1s—F-BANQ (151), F-BBZE (250), F-BCHI (314), F-BDYA (382), and F-BCHY (387).
Air Fret Transimex:	Operated a solitary AAC.1—F-BDYE (404).
SA Air Nolis:	Had two AAC.1s—F-BBYT (240) and F-BBZA (246).
Autrex:	Based in French Indochina, Autrex had two AAC.1s—F-BBYF (226) and F-BBZL (240). Both were lost in crashes.

Avions Bleus:	Owned by Francois Moreau, two AAC.1s were used on charter flights—F-BCHZ (383) and F-BDYB (385).
CATI:	One of a number of small charter companies supporting French forces in Indochina, CATI had only one AAC.1—F-BANQ (151).
Compagnie de Transports Aériens Languedoc Roussillon:	Had three AAC.1s—F-BCAC (283), F-BCAD (284), and F-BCAE (286).
LASO France:	A subsidiary of Air Nolis, LASO had four AAC.1s—F-BBYE (225), F-BBZL (240), F-BBZA (246), and F-BDYD (390).
Service Transatlantique Aérien, Lyons (STA):	Had three AAC.1s—F-BBYA (221), F-BBYB (222), and F-BBYC (223). In addition, c/n 285 was intended for delivery to STA, but was not taken on charge.
Ste Auxiliaire de Navigation Aérienne (SANA):	Along with Socotra, SANA was the largest non-scheduled user of the AAC.1 with eight aircraft—F-BBZF (194), F-BBZG (196), F-BBZI (202), F-BBZJ (203), F-BBZK (204), F-BBZA (246), F-BDYH (401), and F-BDYE (404).
Ste Air Ocean:	Operated three AAC.1s in 1946–47, losing F-BCAA (321) in an accident on 10 October 1946, with F-BBYD (224) and F-BCHE (308) suffering the ignominy of being impounded at Le Bourget on 18 June 1947.
Ste Alpes Province:	Had one AAC.1—F-BCHR (394)
Ste Anonyme de Transport Aérien (Socotra):	Had eight AAC.1s—F-BBZB (247), F-BBZG (248), F-BCAB (282), F-BCHD (311), F-BCHT (380), F-BCHU (386), F-BCHV (391), and F-BDYC (400).
Ste de Transports Aériens Intercontinentaux (TAI):	Had seven AAC.1s in 1946–47—F-BBYK (231), F-BBYL (232), F-BBYM (233), F-BBYN (234), F-BCHA (312), F-BCHG, and (324), and F-BGHX (381).

Great Britain

Somewhat surprisingly, Junkers Ju 52/3ms were operated by British Airlines before, during, and after the war. In early 1937, British Airways (Ltd) acquired two Ju 52/3ms second-hand from the Swedish airline ABA. The Ju 52/3ms were intended for the London–Berlin night-mail services, both being equipped with Lorenz radio blind-landing system. The first, registered as G-AERU, arrived at Gatwick on 29 January 1937, with G-AERX following shortly afterwards. Interestingly, Airports Ltd immediately complained that the 'heavy' Ju 52/3m was damaging the wet area on the eastern side of the runway. With Gatwick not having Lorenz gear fitted, conversion training was transferred to Gravesend, where such gear had already been installed. However, regular maintenance was retained at Gatwick. One additional Ju 52/3m was purchased factory fresh from Junkers in 1939 and registered as G-AFAP.

The outbreak of war meant that the Berlin route was cancelled. British Airways fleet of German airliners was nearly increased when the sole LOT Ju 52/3m, which had been evacuated to Bucharest following the German invasion of Poland, was sold and registered as G-AGAE. However, this particular Ju 52/3m never reached British shores, remaining in Romanian hands.

The trio of British Airways Ju 52/3ms was camouflaged at Whitchurch, with the camouflage pattern allegedly being copied from that of a Fairey Battle single-engined bomber parked at the airfield. At least two of the Ju 52/3ms, G-AERU and G-AERX, were subsequently re-engined with P & W Wasp R-1340-S3H1-G engines and three-blade propellers, which eased maintenance and spares issues considerably (it is unclear if this took place during the autumn of 1939, or in the summer of 1940, following the loss of G-AFAP). Sometime in November 1939, the Ju 52/3ms were ferried to Perth, from where the Nordic services were to recommence. It was intended that the Stavanger–Stockholm service would be maintained by the Ju 52/3ms. The Ju 52/3ms were flown northwards to Scotland under heavy escort by RAF fighters. Services resumed, but on 30 November 1939, one Ju 52/3m, flown by Capt. Geoffrey Mustard, was halfway between Stockholm and Helsinki when a radio report was received that the Finnish capital was being raided by Soviet bombers—Mustard returned to Sweden.

On 1 April 1940, Imperial Airways and British Airways Ltd were merged as British Overseas Airways Corporation (BOAC). The Ju 52/3ms became known as J-Class airliners, with the 'J' indicating Junkers. However, the three Ju 52/3ms were reduced to two within little more than one week.

On 9 April 1940, two of the British Airways Ju 52/3ms were in Scandinavia, G-AFAP in Oslo and G-AERX in Stockholm. With G-AFAP seized by German troops at Oslo Airport, Fornebu, Capt. Ralph 'Lofty' Whitehead and his crew managed to walk across the border to Sweden. The Ju 52/3m was subsequently incorporated into Lufthansa.[6] It is reported that *Oberst* von Winterfeld (once the Station Manager for Lufthansa at Croydon) sent the following message: 'Received from British Airways Ltd one Junkers Ju 52/3m in excellent condition—with thanks. It may be returned to you after the war! [signed] Winterfeld'.[7]

The Ju 52/3m stuck in Sweden was flown back to Britain by Capt. Mustard. With the aeroplane lacking the necessary range, auxiliary fuel tanks were fitted in the cabin. Following take-off, Capt. Mustard brought the Ju 52 to its maximum service ceiling, about 20,000 feet, and the crew—including the British Airways Station Superintendent Maurice Bowden—then took turns to pump fuel from the unpressurised cabin. With no oxygen supply available, this was hard work, being performed in very short intervals. While over the North Sea, they were attacked by a Luftwaffe Dornier Do 18 flying boat. Capt. Mustard had to dive towards the sea, finally reaching British shores at wave-top level.

It was obvious that using the Ju 52/3ms as airliners anywhere on the British Isles would be near suicidal. With Britain facing the risk of German invasion, it was expected that hundreds of Luftwaffe Ju 52/3ms would drop paratroopers over the fields of southern England. As a result, G-AERU and -AERX were overhauled at Gatwick, with one of them being finished in RAF markings, the intent being to produce air-to-air aircraft recognition photographs.

On 4 July 1940, both of the Ju 52/3ms were flown to RAF Speke, Liverpool, from where they were shipped to Lagos, Nigeria. The ownership was formally transferred to British Overseas Airways Corporation (BOAC) on 22 August. Only one of the Ju 52/3ms, G-AERX, entered BOAC service, with G-AERU being used as a source for spares. G-AERX was briefly operated on the Takoradi–Khartoum and Cairo–Khartoum-Lagos routes. On 30 September 1941, both Ju 52/3ms were sold to SABENA for operations in Belgian Congo.

British Pre-war Civilian Ju 52/3ms

Registration	Name	W Nr	P/i	Date Registered	Remarks
G-AERU	*Juno*	5440	SE-AER	19.02.1937	To SABENA as spares, 08.1941

G-AERX	*Jupiter*	5518	SE-AES	07.04.1937	To SABENA in 09.1941. Registered OO-CAP
G-AFAP	*Jason*	5881		28.01.1938	Stranded at Oslo, 09.04.1940. To Luftwaffe
G-AGAE	?	5588	SP-AKX	07.11.1939	Interned at Bucharest, Romania, 09.1939. To LARES

Post-war British Civil Ju 52/3m Operations

It is a relatively unknown fact that a number of former Luftwaffe Ju 52/3ms were rebuilt by Shorts and Harland in the immediate post-war years. By the end of the war, the Douglas C-47s supplied under Lend-Lease had to be returned to US authorities, or purchased using precious US Dollars. With the British aircraft industry having concentrated on combat aeroplanes, few indigenous transport aeroplanes were readily available. One solution suggested by Dennis Handover, Air Adviser to the Ministry of Civil Aviation (MCA), was to use captured stocks of Ju 52/3ms as *ad hoc* airliners. Although BOAC refused this proposal, a number of smaller airlines showed some interest, particularly Railway Air Services. The Ju 52/3m, along with the Fw 200 Condor, Junkers Ju 352, and Siebel Si 204, had already been demonstrated to representatives of the Associated Airways Joint Committee (AAJC) on 28 July 1945.

In the event, Railway Air Services requested the supply of an initial six Ju 52/3ms from RAF stocks for use of domestic routes. However, the MCA requested the transfer of a total of thirty Ju 52/3ms, with another ten for spares recovery. Even though BOAC had refused to contemplate the use of the pre-war Ju 52/3m, another fifty Ju 52/3ms were set aside for possible use by BOAC, just in case the airline's request for fifty-six C-47s fell through.

Locating and refurbishing such a large number of Ju 52/3ms meant that a special RAF unit, the Enemy Aircraft Servicing and Storage Unit (EASSU), was formed on 27 September 1945 at Hamburg Airport, Fuhlsbüttel. Aircraft would then be ferried to 6 MU at RAF Brize Norton for disposal. Short Brothers and Harland at Belfast were contracted to rebuild Ju 52/3ms to civilian standards. Following an engineering assessment of a single airframe, registered as G-AHBP, a contract was made for the rebuild of ten Ju 52/3ms (the eventual cost for rebuilding ten Ju 52/3ms was £155,315, a vast sum considering the eventual low utilisation of the aircraft). Consideration was given to placing a separate contract with Scottish Motor Traction Ltd, however this fell through. The BMW 132Z engines were refurbished by Field Consolidated Aircraft Services Ltd. According to initial plans, each of the Ju 52/3ms was to fly 2,000 hours per annum on average, with the engine time between overhaul (TBO) being estimated at 650 hours.

The Ju 52/3ms could carry twelve passengers with 33 lb of luggage allowed for each passenger. However, various problems meant that the Ju 52/3ms did not enter service until late 1946. Among these problems were lack of exhaust collector rings, locating tyres, and the specific type of engines needed. The Air Registration Board also placed demands for more detailed inspection of each airframe, searching for signs of corrosion, etc. Nevertheless, a Type Certificate for the Ju 52/3m was issued on 10 August 1946. The first Shorts-rebuilt Ju 52/3m, G-AHOC, was delivered to Railway Air Services on 9 November 1946. A further three Ju 52/3ms had arrived by 1 February 1947. On this date, all private airlines were nationalised, with British European Airways taking over the aircraft of RAS.

Services between Croydon and Nutts Corner began in November 1946. On 20 March 1947, the Belfast–London route was inaugurated. The aircraft were also heavily used on various Scottish routes, including the Orkneys and Shetlands. One Ju 52/3m, G-AHOK, was damaged beyond economical repair at Renfrew Airport, Glasgow, on 26 January 1947. Various maintenance issues, including lack of certain spare parts, became a major headache. The high cost of operating the Ju 52/3ms was another factor in the decision to ground the remaining Ju 52/3ms on 10 August 1947. Single flights took place after this date, with the last one apparently being a flight between Glasgow and Liverpool on 31 August. The aircraft were later ferried to Ringway for scrapping. The Ju 52/3ms had lasted some ten months before being withdrawn from use and scrapped.

Post-war British Civilian Ju 52/3ms

Registration	W Nr	SH c/n	P/i	Registered	Cancelled	Remarks
G-AHBP	6750	SH.7C	D-APZX, VM908	07.10.1946	02.02.1948	Scrapped 02.1948 at Castle Bromwich
G-AHOC	501441	SH.16C	VM923	21.05.1946	02.02.1948	Scrapped 02.1948 by British Aluminium Co Ltd, Latchford, Warrington
G-AHOD	131150	SH.10C	VN740	21.05.1946	02.02.1948	*Ibid.*
G-AHOE	?	SH.8C	VN723	21.05.1946	02.02.1948	*Ibid.*
G-AHOF	?	SH.9C	VN729	21.05.1946	02.02.1948	*Ibid.*
G-AHOG	3317	SH.17C	VM979	21.05.1946	02.02.1948	*Ibid.*
G-AHOH	641364	SH.14C	VN746	21.05.1946	02.02.1948	*Ibid.*
G-AHOI	641227	SH.13C	VN744	21.05.1946	02.02.1948	*Ibid.*
G-AHOJ	500138	SH.15C	VN756	21.05.1946	02.02.1948	*Ibid.*
G-AHOK	2998	SH.12C	VN742	21.05.1946	27.01.1947	Dbr 26.01.1947 at Glasgow-Renfrew. Scrapped by Scottish Aviation at Prestwick
G-AHOL	614213	SH.11C	VN741	21.05.1946	02.02.1948	Scrapped 02.1948 by British Aluminium Co Ltd, Latchford, Warrington

British CASA 352s

In 1976, the Warbird Collector Doug Arnold purchased five CASA 352s. Based at Blackbushe, Hampshire, Arnold collected far and wide, obtaining a large number of Second World War-era aeroplanes. All of the CASA 352s were civilian registered, and used for film work in Britain and continental Europe. By the late 1980s, all of the CASAs had been disposed of.

One US-registered CASA 352 (N9012P, EdA serial T.2B-140) was briefly operated during the mid-1980s on the British air-show circuit by the Junkers Ju 52/3m Flight, before being sold in Germany. It is currently on display at Sinsheim.

British Civilian CASA 352s

Registration	C/n	P/i	Registered	Cancelled	Remarks
G-BECL	24	T.2B-212	27.07.1976	19.06.1990	Sold in France. R as F-AZJU
G-BFHD	146	T.2B-255	23.11.1977	21.01.1988	Sold to Lufthansa. Currently preserved at NASM, USA
G-BFHE	164	T.2B-273	13.11.1977	21.01.1988	Sold in Germany
G-BFHF	166	T.2B-275	23.11.1977	28.08.1981	Sold in South Africa
G-BFHG	153	T.2B-262	23.11.1977	27.09.1994	Sold in USA

Greece

Three Ju 52/3ms powered by BMW Hornet engines were ordered by the airline *Elliniki Etairia Enaerion Synhinoion* (EEES). The first aircraft was handed over at Dessau on 30 June 1938, with the other two arriving in Greece in July and August respectively. The Ju 52/3ms were registered as SX-ACF, -ACH, and -ACI respectively on 17 September 1938. The first route outside mainland Greece, to Candia on the Island of Crete, was inaugurated on 13 April 1939. It was hoped that the route, pending permission from the Egyptian government, would be extended to Egypt (which had been requested already in October 1938. In all likelihood, this was connected to the purchase of the Ju 52/3ms).

Initially, the route was flown three times a week. However, owing to an increase of passengers and mail, a fourth weekly flight was added on 15 May. Following the Italian invasion of Greece in October 1940, all three Ju 52/3ms were requisitioned by the Hellenic air arm. The airline did not close shop, however. A new twenty-year concession was received in May 1941 (when Greece was under German occupation). On 13 June 1941, Lufthansa took a 51 per cent controlling interest of EEES. With the collapse of the Greek economy in 1941–42, Lufthansa offered EEES a Ju 52/3m in exchange for the remaining shares of the company. If delivered, the Ju 52/3m would be used to operate a domestic air service—this did not happen, and EEES did not return to the airline business after the end of the war.

Post-war

Following the German surrender, the Greek government requested the supply of Ju 52/3ms from RAF stocks. None were received, though.

In 1949, one Ju 52/3m was acquired from Sweden by Hellenic Airlines (*Aeropirokai Metaphorai Ellados* (AME)). In July 1951, AME merged with other Greek airlines, resulting in the formation of National Hellenic Airlines. The Ju 52/3m, however, had been withdrawn from use before the merger.

Greek Civil Ju 52/3ms

Registration	*Werk Nummer*	Date Registered	Remarks
SX-ACF	5984	17.09.1938	To Hellenic Air Force 02.11.1940
SX-ACH	6004	17.09.1938	To Hellenic Air Force 02.11.1940
SX-ACI	6025	17.09.1938	To Hellenic Air Force 02.11.1940
SX-CDA	5614	1949	P/i SE-AFD

Hungary

Formed on 19 November 1922, Magyar Légiforgalmi R. T. (better known as MALÉRT) purchased two Ju 52/3ms in 1936. That same year, a Ju 52/3m was supplied to the Hungarian Air Ministry, being registered as HA-DUR (the Hungarian Air Ministry's Type Designator for the Ju 52/3m was 10.01). The aircraft was predominantly used by the country's leader, Miklos Horthy. The Air Ministry Ju 52/3m was eventually transferred to MALÉRT in 1938, being re-registered as HA-JUC. MALÉRT operated daily services with their three Ju 52/3ms on the Budapest–Krakow–Warsaw, Budapest–Prague, and Budapest–Vienna routes, pooling resources with Lufthansa. Additional aircraft were delivered in 1939–41.

Following the German invasion of the USSR on 22 June 1941, three Ju 52/3ms were transferred to the RHAF. In 1942, the route Budapest–Bratislava–Vienna–Breslau–Berlin was established, being served exclusively between July and September 1942, by MALÉRT Ju 52/3ms. However, with Hungarian forces being pushed back on the Eastern Front, MALÉRT reduced their regular scheduled flights. By late 1942, the remaining Ju 52/3ms had been transferred to the Royal Hungarian Air Force.

Hungarian Civil Ju 52/3ms

Registration	W Nr	Name	P/i	Date Registered	Remarks
HA-DUR	5600, g4e	*Kiss Josef*		06.1936	Re-registered HA-JUC 1938
HA-JEA	3288, g9e		TH+UT	1942	To RHAF 09.1942.
HA-JUA	5523, g3e	*Kaszala Karoly*		06.1936	W/o Nagy-Varad 18.01.1941 RHAF s/n U.861 ntu
HA-JUB	5580, g3e	*Gergye Jozsef*		06.1936	To RHAF 1942.
HA-JUC	5600, g4e	*Kiss Jozsef*	HA-DUR	1938	To RHAF 1941, s/n S.101, to U.863 in 06.1942.
HA-JUD	6360, g3e	*Endresz Gyorgy*		1939	Re-registered HA-JUF 1940

A MALÉRT Ju 52/3m, seen here following an 'argument' with a British Airways Lockheed 10. (*Bernd Huckenbeck*)

HA-JUE	6585, g7e	*Szentkiralyi Dezso*		1939	To RHAF 06.1942, s/n U.864
HA-JUF	6360, g3e	*Endresz Gyorgy*	HA-JUD	1940	To RHAF 06.1942, s/n U.865
HA-JUG	7005, g7e	*Szent-Istvany Dezso*		1941	To RHAF 06.1942, s/n U.866
HA-JUH	7041, g3e	*Dulo Mihaly*		1941	To RHAF 06.1942, s/n U.867

Italy

The Italian airline Ala Littoria S.A. operated on a multitude of routes in Europe, North Africa, and South America. Although predominantly Italian-built aeroplanes were used, three Ju 52/3ms powered by Piaggio XR engines were ordered, with the first arriving in early 1935. These were registered as I-BAUS, I-BEZI, and I-BIZI.

One Ju 52/3m, powered by BMW engines, was obtained second-hand in late 1938 and registered as I-ABJZ. It remained in service for four months only before being returned to Lufthansa. During its brief sojourn with Ala Littoria, I-ABJZ was based at Venice.

On 4 December 1939, I-BAUS was lost in a crash near Formia, with nineteen being killed. In June 1940, Italy declared war against France and Britain. This resulted in four Ala Littoria Ju 52/3ms being transferred to the *Regia Aeronautica*. One Ala Littoria Ju 52/3m, registered in August 1940 as I-BERO, was powered by Alfa 126 engines. By the early 1940s, no Ju 52/3ms remained in service.

Italian Civil Ju 52/3ms

Registration	W Nr	P/i	Registered	Cancelled	Remarks
I-ABJZ	5942	D-AUJG	09.11.1938	03.1939	To D-AUJG 02.1939, w/o 04.08.1939 between Barcelona and Madrid
I-BALI	6779		30.01.1940		Powered by BMW 132 engines. Fate unknown
I-BAUS	4063		21.01.1935	12.1939	Crashed into Mont Maranola, near Formia, 30.04.1939, nineteen killed
I-BERO	6803		08.1940	?	To RA M.M.60411, to D-ASPE, to Lw AS+PE
I-BEZI	4062		03.01.1935	06.1939	Possibly sold to DLH in 1937
I-BIOS	6710		06.1940	?	To RA M.M.60412, to D-ASPI, to Lw AS+PI
I-BIZI	4064		18.02.1935	?	To RA M.M.60410, to D-AIAO, to Lw AI+AO
I-BOAN	6765	RA M.M.60413	09.1940	?	To D-AIAT, to Lw AI+AT

Lebanon

Formed at Beirut in April 1945, the airline Compagnie Générale de Transports (CGT) received three former Air France AAC.1s the following year. The initial two AAC.1s were registered as L-RAMT

An Ala Littoria Ju 52/3m, I-ABJZ. (*Rob Mulder*)

Ju 52/3m I-BIZI of Ala Littoria. This, as well as several other Ala Littoria Ju 52/3ms, was powered by Piaggio XR engines. Note the different shape of the engine cowlings. (*Gregory Alegi*)

and L-RAMY in February 1946, with the registrations being changed three months later to LR-AAJ and LR-AAI respectively. The third aircraft, LR-AAC, arrived in June 1946. CGT operated on regular routes to Khartoum, Kano, Lagos, Tunis, Casablanca, Dakar, and Cairo. By 1947, the AAC.1s had been largely replaced by Douglas DC-3s. In April 1948, one of the AAC.1s was sold to Yemen. Apparently, it did not arrive in Yemen, possibly being requisitioned by the Syrian Air Force. The remaining two AAC.1s were definitely transferred to the Syrian Air Force on 9 July 1949.

Lebanese Civil AAC.1s

Registration	C/n	P/i	Date registered	Remarks
LR-AAC	73	F-BAKU	06.1946	Sold to Yemen on 13.04.1948. Not delivered, possibly top SyrAF
L-RAMT	109	F-BAMT	02.1946	Re-registered as LR-AAI on 16.05.1946. To SyrAF 09.07.1949
L-RAMY	113	F-BAMY	02.1946	Re-registered as LR-AAJ on 16.05.1946. To SyrAF 09.07.1949

Mozambique

The Portuguese colony of Mozambique may seem an unlikely operator of the Ju 52/3m. However, three Ju 52/3ms were delivered in 1938 to the airline DETA (*Divisao de Exploracao dos Transportes Aereos*) Mozambique and used for internal routes, as well as Johannesburg in South Africa and Tananarive in Madagascar. The three aircraft were registered as CR-AAJ (W Nr 5962, named *Lourenco Marques*), -AAK (W Nr 5967, named *Quelimane*), and -AAL (W Nr 5973, named *Nampula*) respectively. CR-AAJ crashed on 12 February 1950, with the other two remaining in service until 1957. The hulks of -AAK and -AAL were still present at Maputo until the early 1980s, but had gone by 1985.

Netherlands

In 1946, a small number of Ju 52/3ms were transferred to the *Rijksluchtvaartschool* (RLS). The RLS was responsible for training commercial pilots for the national airline KLM. The Ju 52/3ms were used as multi-engined trainers. Late in its service life, PH-UBA was used as an engine test bed. It was withdrawn from use in 1950, then being transferred to Avio-Diepen NV for static display before being scrapped in January 1951.

Dutch Civil Ju 52/3ms

Registration	Name	W Nr	P/i	Registered	Cancelled	Remarks
PH-UBA	*Opa*	5717 (5715 also quoted)	VM985	21.06.1946	24.01.1951	Built by Weser Scrapped
PH-UBB	*Ome Keesje*	3428	3K+LN, VM927	13.07.1946	30.09.1949	Built by ATG, Leipzig Scrapped
PH-UBC	*Oma*	05210078	VN743	14.10.1946	24.01.1951	Built by ATG, Leipzig Scrapped

CR-AAK of DETA seen at Johannesburg shortly before the war. (*Yolanda Meyer*)

One of the three RLS Ju 52/3ms, PH-UBC. (*Thijs Postma*)

A fourth Ju 52/3m, serial VM907, was handed over by the RAF, having suffered damage at Eindhoven on 25 November 1945. It was used as spares.

Nicaragua

The much-travelled *Aconcagua* (W Nr 5283) ended its working life in Nicaragua. In early 1948, the Ju 52/3m was registered as AN-ACS for TACA de Nicaragua. However, its service was relatively brief, suffering irreparable damage in a landing accident at Monte Carmelo later the same year. The old Junkers was then broken up by using a bulldozer. As noted by US aviation historian Dan Hagedorn, *Aconcagua* had been operated by no less than six airlines, two air forces and a US government agency, possibly a record for any Ju 52/3m.

Norway

In 1935, Norwegian Airlines (*Det Norske Luftfartselskap AS* (DNL)), received a ten-year concession covering all regular air traffic in Norway. The first route, from Oslo to Bergen, was established in May 1935, being flown with a Ju 52/3m leased from Lufthansa. Later the same year, the Ju 52/3m was bought, being registered as LN-KAE. It also received the name *Havørn* (*Sea Eagle*). While on a route proving flight to the northern town of Tromsø on 16 June 1936, LN-DAE crashed into mountains near Lindås, all seven aboard being killed. Nevertheless, a further three Ju 52/3ms were acquired by DNL in 1936 (two) and 1937 (one), being registered as LN-DAF, -DAH, and -DAI.

Initially, the Ju 52/3ms operated solely on floats. Following the opening of the Oslo Airport, Fornebu, in June 1939, a land route between Oslo, Gothenburg, and Copenhagen was inaugurated. This route was flown by LN-DAF until 8 April 1940. One Ju 52/3m, LN-DAI, was leased to the Norwegian Navy in early 1940, with this and the other two Ju 52/3ms being captured by German forces on 9 April. However, two of these Ju 52/3ms would make an unexpected return to DNL following the end of the war.

Post-war DNL Operations

Following the German surrender, many Ju 52/3ms were left in Norway. DNL was reformed using refurbished Ju 52/3ms. Following the incorporation of DNL into Scandinavian Airlines System (SAS), two Ju 52/3ms continued in service until the mid-1950s on the coastal routes in northern Norway.

Following the German surrender, a number of Ju 52/3ms were handed over to DNL, including the former LN-DAF and -DAH. Both were refurbished at Horten, with the latter serving briefly with the Norwegian Air Force. Formal ownership was transferred to DNL in 1946, with W Nr 5489 being re-registered as LN-KAF on 18 May 1946, and it was named *Askeladden*. The fuselage was subsequently found to suffer from serious corrosion problems, which resulted in it being replaced in 1947 with the fuselage of a former Luftwaffe Ju 52/3m, W Nr 130714.

On 27 May 1946, the Trondheim–Tromsø service was inaugurated by DNL, being extended to Kirkenes on 13 August. Both of the Ju 52/3ms had been purchased by DNL pre-war, and were

LN-DAI of DNL. (*Mats Averkvist*)

registered as LN-KAF (W Nr 5429) and LN-DAH (W Nr 5489) respectively. On 12 April 1947, a second DNL Ju 52/3m, LN-KAG, was placed on the civilian register and named *Vesslefrikk*. When DNL became part of Scandinavian Airlines System (SAS), both LN-KAF and -KAG were exclusively used on the routes in northern Norway. Interestingly, although both operated in full SAS colours, neither was given a Viking name, as carried by virtually every other SAS airliner.

Following the expiration of its Certificate of Airworthiness in September 1955, LN-KAF was sold to a private individual for a sum of 3,000 NOK. It was briefly used as a plaything at a children's day-care centre near Stavanger Airport, Sola. The second Ju 52/3m was offered to the Norwegian Technical Museum, but rejected due to lack of storage space. Instead, it was sold to Ecuador. Today, this historic Ju 52/3m is airworthy with Lufthansa.

Norwegian Civil Ju 52/3ms

Registration	Name	W Nr	P/i	Date of Registry	Date of Cancellation	Remarks
LN-DAE	*Havørn*	4077	D-ANOP, (LN-DBE) ntu	07.06.1935	24.06.1936	W/o Lihesten, 16.06.1936
LN-DAF	*Najaden*	5429	D-AKIY	06.06.1936	19.08.1940	Taken over by Lw 09.04.1940. R D-AKIY, then to LN-KAG post-war

LN-DAH	*Falken*	5489	D-AQUI	01.07.1936	15.06.1940	Taken over by Lw 09.04.1940. R D-AQUI, then to LN-KAF post-war, R HC-ABS, R N130LW, R N52JU, R D-CDLH
LN-DAI	*Hauken*	5751		13.09.1937	04.02.1940	To Norwegian Navy, s/n F.400. Taken over by Lw 09.04.1940. To DLH, R D-AQUB. W/o 01.03.1943 at Hommelvik
LN-DBE		4077	D-ANOP	Res 10.05.1935, ntu		R as LN-DAE
LN-KAB	*Tyrihans*	7229	Lw PI+xx	30.03.1946	25.05.1948	Wfu, scrapped
LN-KAD	*Per*	130712	Lw PI+xx	08.06.1946	28.11.1948	Wfu, scrapped Stavanger-Sola.
LN-KAE	*Pål*	130704	Lw PI+xx	27.06.1946	28.11.1950	Wfu, scrapped
LN-KAF	*Askeladden*	5489 130714	D-AQUI, LN-DAH,D-AQUI	18.05.1946	25.10.1956	Received fuselage from Lw W Nr 130714. Sold as HC-ABS. Currently flying with DLH as 'D-AQUI'
LN-KAG	Veslefrikk	5429	D-AKIY, LN-DAF, D-AKIY	12.04.19 47	02.07.1956	Wfu, scrapped
LN-KAH	Ulabrand	7215	Lw PI+xx	Res 1946, ntu		Std Stavanger-Sola until 1956, scrapped
LN-KAI	Peik	3257	Lw 8A+AK	17.07.1946	25.05.1948	Wfu, scrapped
LN-KAL		130714	Lw PI+xx	Res 08.05.1946, ntu		Fuselage to LN-KAF
LN-KAM				Res 1946, ntu	1950	
LN-KAN				Res 1946, ntu	1950	
LN-LAB	?	6048	Lw N9+DA, RNoAF 6048	13.04.1946	03.08.1946	W/o 22.05.1946 at Fornebu

Above left: Hugo Junkers (1859–1935). (*Arlanda Civil Aviation Collection*)

Above right: Air Transport was in its infancy during the early 1930s. (*Arlanda Civil Aviation Collection*)

Below: A magnificent view of D-2201 over the Alps. (*Rob Mulder*)

Above: Two of Lufthansa's Ju 52/3ms were experimentally fitted with Jumo 205 diesel engines. Both of the Ju 52/3ms, D-AJYR and D-AQAR, were soon withdrawn from use. (*Rob Mulder*)

Below: Excited and happy passengers seen exiting D-ANYF in 1935. (*Rob Mulder*)

D-AZAZ of Lufthansa: note the enormous wheel spats. (*Mats Averkvist*)

A Luftwaffe Ju 52/3m during the Polish campaign. Note the short-lived WL+ code. (*NMUSAF*)

One of the three Ju 52/3ms, D-TABX, used to deliver supplies to isolated areas along the North Sea coast in early 1940. (NMUSAF)

Three Ju 52/3ms made forced-landings on the Den Haag highway. (*Thijs Postma*)

Two Ju 52/3ms seen after overrunning the runway at Maleme. (*Author's Collection*)

Most of the air supply flights on the Eastern Front were flown with the bare minimum of ground support equipment. (*Author's Collection*)

Flying at low altitude and keeping a tight formation provided at least some protection from roving Allied fighters. (*Author's Collection*)

A Ju 52/3m gl4e seen late in the war. For many *Transportverbände* pilots, flying operational sorties in poor weather conditions had at least one advantage—the enemy was affected by the weather as well, resulting in reduced risk for fighter opposition. (*Author's Collection*)

The five Ju 52/3ms of Aeroposta lined up. (*Michael Magnusson*)

VH-BUV seen at Bromma airport, Sweden, before embarking on the flight to Australia. (*Hans Boëthius via Swedish Aviation Historical Society*)

PP-CBB seen by the floating dock in Buenos Aires harbour. (*Author's Collection*)

A rare air-to-air view of CF-ARM. Note the oversized skis. (*Larry Milberry*)

A nice view of OY-DFU, used on the Copenhagen–Bornholm route after the war. (*Tom Weihe*)

AAC.1 F-BALN was delivered to Air France in July 1945, subsequently being operated by CPEM for air-mail services. It was transferred to *Armée de l'Air* in April 1949. (*Collection Jean Delmas via Jean-Pierre Dubois*)

HA-JUA of MALÉRT. A DC-3 of KLM can be seen in the background. (*Rob Mulder*)

The first three SAA Ju 52/3ms seen during the delivery flight. (*Yolanda Meyer*)

Ju 52/3m ZS-AJC of SAA seen before the war. (*Yolanda Meyer*)

EC-CAL of Iberia. (*Luis Gonzalez Pavón*)

A rare colour photo of SE-AFB. This and the other Ju 52/3ms of ABA were finished in overall neutral orange following outbreak of war. (*Arlanda Civil Aviation Collection*)

Above: In common with other *Flugwaffe* transport and liaison aeroplanes, the Ju 52/3ms were issued with civilian registrations to permit flights outside Switzerland. (*Bernd Huckenbeck*)

Below: CAUSA's Ju 52/3ms operated on 'high-density' (as they are currently known) flights between Montevideo and Buenos Aires, being fitted out to carry twenty-eight passengers. (*Michael Magnusson*)

FAA Ju 52/3m s/n T-160 seen during a parade. Note the Curtiss Hawk fighter and Northrop 8A-2 light bomber in the background. (*Michael Magnusson*)

Above left: Czech Air Force Ju 52/3m s/n D-5 seen at Bromma Airport, Sweden, soon after the end of the war. (*Mats Averkvist*)

Above right: An AAC.1 Toucan of GT 1/64 *Béarn* seen flying over Hanoi in February 1950. (*Albert Grandolini*)

Above left: Dropping paratroopers during operations in northern Vietnam. (*Albert Grandolini*)

Above right: Refuelling a Toucan of GRSA 76 on a landing strip in the Saharan Desert. (*Albert Grandolini*)

Ju 52/3m code 22-67 seen following the end of the Civil War. (*Luis Gonzalez Pávon*)

The red-and-white neutrality markings were introduced in 1944 following an incident where USAAF P-51 Mustangs clashed with Swiss Bf 109Es. (*Bernd Huckenbeck*)

LQ-ZBD is the oldest Ju 52/3m to survive—it is seen here in 2008. The aircraft in the background is an Avro Lincoln bomber. (*Michael Magnusson*)

F-AZJU is a regular performer at La Ferté-Alais. (*Miro Patocka*)

Marked as 'D-AZAW' and on display at Berlin's *Museum für Verkehr und Technik*, this particular Ju 52/3m has a very diverse and interesting history. (*Björn Zereba*)

A CASA 352, marked as 'D-ANOY', can be seen at Munich airport. (*Hennies Werner*)

CASA 352 masquerading as Ju 52/3m 'G-AFAP' at the RAF Museum, Cosford. (*Per Björkner, Arlanda Civil Aviation Collection*)

One of the Ju 52/3ms recovered from Lake Hartvikvannet, CA+JY has been restored for the Norwegian Air Force Museum at Gardermoen. (*Mats Averkvist*)

Above: AAC.1 s/n 7208 seen at Surčin in the early 2000s. (*Author's Collection*)

Opposite: ZS-UYU seen over the Hartbeespoort dam. (*Sonja Grünbauer*)

ZS-AFA

Above: The second CASA 352 at Cuatro Vientos, s/n T.2B-254, was originally delivered to the EdA in 1954. (*Luis Gonzalez Pavón*)

Below: The rare sight of six airworthy Ju 52/3ms and CASA 352s in formation. (*Bernd Huckenbeck/Ju Air*)

Peru

The German-sponsored airline Deutsche Lufthansa Sucursal Peru began operations on 24 May 1938. Based at Lima, Sucursal Peru operated on the Lima–Arequipa–Puno–Arica route, which continued to the Bolivian capital La Paz—plans for a connecting route to Ecuador never reached fruition. In all four Ju 52/3ms were registered to Sucursal Peru. Due to American pressure, the Peruvian government withdrew the operating permit, resulting in the suspension of operations on 31 March 1941. Two aircraft were seized by the Peruvian government on 31 March, both being transferred to the Peruvian Aeronautical Corps (*Cuerpo Aeronáutico del Perú* (CAP)).

Peruvian Civilian Ju 52/3ms

Registration	W Nr	Name	P/i	Remarks
OA-HHA	5060	*Huascarán*	D-AMIT	Seized by Peruvian gvt 31.03.1941
OA-HHB	5043	*Misti*	D-ARYS	W/o 26.06.1938
OA-HHC	5272	*Huandoy*	D-AMIQ	Seized by Peruvian gvt 31.03.1941
OA-HHD	5283	*Aconcagua*	D-AENF	Re-registered D-AENF. Leased to SEDTA 29.11.1940

Poland

One solitary Ju 52/3m was purchased by Polish Airlines (*Polskie Linie Lotnicze* (LOT)), and registered on 12 November 1936 as SP-AKX. Uniquely, it was powered by three 725-hp Bristol Pegasus VI engines. Following outbreak of war on 1 September 1939, SP-AKX was flown south to Romania on 12 September, along with several other LOT airliners. Following arrival at Bucharest, LOT attempted to sell the aeroplanes to Imperial Airways. Although registered as G-AGAE, the Junkers Ju 52/3m never reached British shores, in no small part due to the intervention by the German embassy in Bucharest. Instead, it was taken over by Romanian authorities.

Polish Civil Ju 52/3m

Registered	W Nr	Date registered	Remarks
SP-AKX	5588	12.11.1936	Flown to Romania 12.09.1939

Portugal

One AAC.1 was briefly operated by Aero Portuguesa Lda., Lisbon. Originally formed as a subsidiary of Air France, a 50 per cent share was sold to Portuguese shipping interests in 1943. The AAC.1 was registered as CS-ADA (c/n 148), being returned to Air France on 27 October 1947.

Romania

The first Ju 52/3m to arrive in Romania was an early production aircraft, W Nr 4016. Powered by Hispano-Suiza engines, the Ju 52/3m ba was delivered in 1932 as a VIP aircraft for Prince George Bibescu (1880–1941), then President of the Fédération Aéronautique Internationale (FAI). The Ju 52/3m replaced a Ford Trimotor, which had been lost following a mid-air collision with a vulture over India. The Ju 52/3m was registered as CV-FAI in April 1932.

The Ju 52/3m was later re-registered as YR-ABF, eventually being transferred to the Romanian state airline, LARES, with the non-standard engines being exchanged for BMW Hornet units. The Ju 52/3m was damaged beyond repair at Barzagio-Dobruja, near Bucharest, on 23 May 1940. In September 1939, no less than fourteen Polish airliners from LOT arrived in Romania. Included among these was a solitary Ju 52/3m, SP-AKX, which arrived at Băneasa on 12 September. Despite attempts by LOT to transfer the fleet of airliners to Britain, all of them were duly impounded by the Romanian government. Interestingly, the Ju 52/3m was not formally registered as YR-ALK until 29 April 1942, as YR-ALK. On 16 October 1940, LARES was formally incorporated into the Romanian Air Force. In February 1941, only one Ju 52/3m remained in LARES service. In June 1941, the sole remaining LARES Ju 52/3m was transferred to the Romanian AF.

A small number of Ju 52/3ms were delivered to LARES in 1944, with additional aircraft being transferred from the Romanian Air Force. When Romania became a Communist-led state on 31 December 1947, the remaining Ju 52/3ms were transferred to AVSAP. By the late 1950s, the last Ju 52/3ms had been withdrawn from use.

The VIP interior of CV-FAI. (*Swedish Aviation Historical Society*)

Romanian Civilian Ju 52/3ms

Registration	W Nr	P/i	Date registered	Date Cancelled	Remarks
CV-FAI	4016		04.1932		Re-registered YR-ABF
YR-ALK	3588	SP-AKX, G-AGAE ntu	29.04.1942		
YR-ALL	131406		12.07.1944		
YR-ALM	131407		12.07.1944		Wfu 1958
YR-ALN	7363	Lw NJ+KL			
YR-ALO	131408		17.11.1944	23.12.1958	Wfu 02.1958
YR-ALP(1)	6428	Romanian AF	29.03.1952 *	25.02.1958	
YR-LIV	35210		25.04.1953	23.12.1958	
YR-PTV			25.10.1949		

*Unconfirmed

South Africa

The national airline South African Airways (SAA) operated a fairly large number of Ju 52/3ms before the war. In 1940, all of the surviving aircraft were requisitioned by the South African Air Force (SAAF). None returned to civilian use after the end of the war.

In 1933, the privately owned airline Union Airways Ltd, which was supported by the local Junkers representative, Herr F. Hoepfner, ordered three Ju 52/3ms. However, before the aircraft were delivered, Union Airways was taken over on 1 February 1934 by the South African Railways and Harbours Administration. The airline was suffering from a poor financial situation, and a number of highly publicised crashes. As a result, the state-controlled South African Airways was formed at Durban. The order for three Ju 52/3ms was increased to four, with the first aircraft arriving on 1 June 1935, and, the following month, the SAA headquarters moved to Rand Airport, Germiston. The arrival of the Ju 52/3ms meant an increase of domestic routes. Apart from carrying passengers, the Ju 52/3ms were occasionally used to deliver newspapers to isolated farms. Such deliveries were greatly assisted by the fact that the cabin windows could be opened in flight.

The four Ju 52/3ms were powered by BMW 132as engines, driving two-bladed fixed pitch propellers. Engine synchronisation was performed by means of mirrors fixed to the outer engine cowlings. The process of synchronisation involved coordinating the outer engines to the centre engine by observing the strobe effect visible in the mirrors!

Regular passenger services expanded to such an extent that a further ten Ju 52/3ms were ordered, with an additional aircraft being leased. These Ju 52/3ms were all powered by Pratt & Whitney Hornet engines, driving three-bladed propellers (the delegations from SAA and Swedish Airlines visited Dessau at the same time; hence the identical requirement in changing the design of the outer engine mounts. These, along with the three aircraft operated by DETA, were the only Ju 52/3ms having this configuration). The initial three aircraft were all returned to Junkers as part-payment, with one—W Nr 4061—subsequently being sold to Bolivia. When war erupted in September 1939,

SAA maintained a reduced number of routes. In the event, all of the remaining Ju 52/3ms were requisitioned by the South African Air Force (SAAF) for use as troop transports.

In the early 1980s, CASA 352 was acquired from Great Britain and restored to airworthiness. Registered as ZS-UYU, this aircraft is still flying. The history of this particular CASA 352 is related elsewhere.

South African Civil Ju 52/3ms

W Nr	Variant	Registration	Name	Date Registered	Cancelled	Remarks
4058	Mge	ZS-AFA	*Jan van Riebeeck*	11.1934	10.03.1938	Sold to VARIG, R PP-VAL
4059	Mge	ZS-AFB	*Lord Charles Somerset*	12.1934	30.06.1938	R D-ACBO, later M-CABO
4060	Mge	ZS-AFC	*Simon van der Stel*	11.1934	30.06.1938	R D-AGFD, R M-CABU, R EC-AAI
4061 *	Mge	ZS-AFD	*Sir Benjamin D'Urban*	04.1935	27.06.1947	To SAAF, s/n 660
5773	Sai	ZS-AJF	*Andries Pretorius*, later *Earl of Caledon*	07.1937	27.06.1947	To SAAF s/n 661
5789	Sea	ZS-AJG	*Piet Retief*	10.1937	27.06.1947	To SAAF s/n 662
5792	Sea	ZS-AJH	*Erasmus Smit*	10.1937	27.06.1947	To SAAF s/n 663
5896	Sea	ZS-AJI	*Major Warden*, later *President Burgers*	09.1937	27.06.1947	To SAAF s/n 664
5899	Sai	ZS-AJJ	*Sir George Grey*	11.1937	27.06.1947	To SAAF s/n 556
5351	Sai	ZS-AKY	*Earl of Caledon*	06.1937		Crashed Rand apt 14.06.1937
5893	Sai	ZS-ALO	*William Adriaan van der Steel*	01.1938	30.06.1947	To SAAF s/n 666
5902	Sai	ZS-ALP	*Louis Botha*	01.1938	30.07.1947	To SAAF s/n 667
5905	Sai	ZS-ALR	*Paul Kruger*	03.1938	30.07.1947	To SAAF s/n 668
5908	Sai	ZS-ALS	*Sir Harry Smith*, later *Thomas Halstead*	01.1938	30.07.1947	To SAAF s/n 669
5911	Sai	ZS-ALU	*Sir Peregrine Maitland*	04.1938	30.07.1947	To SAAF s/n 670

Spain

Formed on 28 June 1927 as Sociedad Iberica de Aviación (Iberia), with a Spanish business consortium holding a majority stake of 51 per cent and the German airline DLH 49 per cent, operations began on 15 December 1927 and ended less than a year later. The Spanish government was favouring a unified airline, which was eventually formed in 1929. However, Iberia remained a legal entity, with services being reintroduced by the rebel, nationalist side in August 1937, with Ju 52/3ms being supplied by Germany. In the first quarter of 1938, the Salamanca–Seville–Tétouan route was inaugurated. Following the end of the Civil War, seven of the Ju 52/3ms were purchased by Iberia on 1 July 1939. The German minority share was acquired by the Spanish government on 7 June 1940, with Iberia being nationalised on 14 August 1943. Four additional Ju 52/3ms were purchased by Iberia in 1941–42.

The Civil War and subsequent outbreak of the Second World War meant a reduction of Iberia routes. However, destinations in both France and Germany were included in the route network, as well as Tétouan and the Canary Islands. The war meant increased risks for civilian airliners, with one Iberia Ju 52/3m, EC-ABD, being destroyed in an Allied bomb raid against Stuttgart on 5 September 1944. A replacement aircraft was eventually received from Lufthansa stocks.

Post-Second World War, three Ju 52/3ms were acquired from Sweden, being used on the route to Spanish Equatorial Africa. The Ju 52/3m remained in Iberia service until the summer of 1961, when the last two, EC-ADO and -ADP were withdrawn from use. No CASA 352s ever served in civilian guise in Spain.

Spanish Civilian Ju 52/3ms

Registration	Werk Nr	Delivered	Name	P/i	Remarks
M-CABA	5854	16.08.1937	*Sanjurjo*	D-AUKE	Lost off Gibraltar. Cancelled 18.12.1939
M-CABE	5851	26.10.1937	*La Cierva*	D-ALUA	Re-registered EC-AAF. W/o Las Palmas. Cancelled 31.12.1945
M-CABO	4059	13.04.1938	*Mola*	ZS-AFB, D-ACBO	W/o Cabezabellosa. Cancelled 11.03.1938.

EC-ADO was one of three Ju 52/3m acquired from Sweden in the late 1940s. (*Luis Gonzalez Pávon*)

M-CABU	4060	10.1938	*Haya*	ZS-AFC, D-AGFD	Re-registered EC-AAI, *Guadiana*. Cancelled 19.10.1950
M-CABY	5098	02.1939	*Ebro*	D-AKYS	Re-registered EC-AAH. Cancelled 19.10.1950
M-CABO	5022	03.1939	*Tajo*	D-AXUT	Re-registered EC-EC-AAG. W/o 31.12.1945
M-CABB	4073	06.1939	*Morato/ Guadalquivir*	D-AVUL	Re-registered EC-AAJ. W/o 01.03.1941
M-CABC	4076	06.1939	*Duero*	OE-LAK, D-AJAT	Re-registered EC-AAK. Cancelled 19.10.1950
EC-AAL	7017	17.03.1941	*Eugenio Gross*	D-AOAW	Cancelled 19.10.1950
EC-CAJ	7053	10.09.1941		D-AQIJ	Re-registered EC-ABD. Destroyed on ground 05.09.1944 at Stuttgart in Allied bomb raid. Cancelled 05.09.1944
EC-CAK	7196	02.01.1942		D-AXFL, CJ+KA	Re-registered EC-ABE. Cancelled 19.06.1943. Transferred to EdA, s/n 22-109
EC-CAN	7220	02.01.1942		D-AZAW	Re-registered EC-ABF. Cancelled 12.05.1943. Transferred to EdA, s/n 22-108
EC-CAL	6725	11.01.1943		JB+NE	Re-registered EC-AAU. Cancelled 19.10.1950
EC-DAN	6015	17.11.1945		D-AMYM 22-105	Cancelled 28.11.1950
EC-DAM	5386	01.05.1946		D-ASUP 22-64	Cancelled 20.06.1955
EC-ADO	5620	02.10.1948		SE-AFB	Cancelled 06.07.1961
EC-ADP	5646	26.10.1948		SE-AFD	Cancelled 06.07.1961
EC-ADQ	5633	10.03.1949		SE-AFC	W/o Yaoundé, Cameroon. Cancelled 30.07.1953

Sweden

Formed in 1924, AB Aerotransport (ABA) had strong connections with Junkers. Initially operating a number of Junkers F.13s, the larger three-engined Junkers G 24 later joined the airline. The first Ju 52/3m arrived in the summer of 1932, and was registered as SE-ADR and named *Södermanland*. The Ju 52/3m was operated on floats on the Stockholm–Helsinki and Stockholm–Reval (i.e. Tallinn) routes during summertime and as a landplane on the Malmö–Amsterdam route during winters. It was also used on the Stockholm–Mariehamn–Turku and Stockholm–Visby routes.

During a visit to Sweden by the Prince of Wales and his brother George in October 1932, SE-ADR was used as a VIP transport. On 5 October, the British royal family were flown from Stockholm to Gävle, where the Sandvik Iron Works and Korsnäs Saw Mill were visited. The Prince of Wales

A picture of the first ABA Ju 52/3m, SE-ADR. (*Arlanda Civil Aviation Collection*)

later commented that the trip had been 'splendid', adding that he was 'astonished that the machine was so well built and isolated, that despite the prevailing cold weather, the passenger cabin felt rather too hot'.[8]

In late 1932, OH-ALK was leased from Aero O/Y, and operated on the Amsterdam route after 17 March 1933. During the winter of 1934–35, OH-ALK was again leased, operating on the Malmö–Hamburg–Berlin route between 22 October 1934 and 2 January 1935.

In 1936, two additional Ju 52/3ms arrived. Registered as SE-AER and -AES, they were briefly operated on the continental air-mail routes, linking Sweden with Germany and Britain. Both were soon sold to British Airways Ltd.

In mid-1936, the fitting of more powerful Bristol Mercury VII engines (which were built under licence by NOHAB) instead of the Wasp S3H1-G was suggested. The Mercury VII had a maximum power output of 685 hp, compared to 550 hp of the Wasp. It was estimated that by using Mercury VII engines, the overall performance characteristics of the Ju 52/3m would increase. This included a 20-kph increase of speed to 305 kph. However, as the Mercury VII was both larger and heavier, the load-carrying capability would be reduced from 1,820 kg to 1,410 kg, with rising operating costs. As a result, the ABA Ju 52/3ms were not powered by Mercury VII engines.

Four additional Ju 52/3ms were delivered in 1936–37, being registered as SE-AFA, -AFB, -AFC, and -AFD. Even though they were not powered by Mercury VII engines, there was one major difference from standard Ju 52/3ms. This involved the repositioning of the outer engines. Instead of the usual incidence, the engines were located in the direct angle of the direction of flight, which meant a reduced torque effect. Interestingly, a South African Airways delegation visiting Dessau at the same time also ordered Ju 52/3ms with this engine configuration. Along with SE-ADR, these Ju 52/3ms remained in ABA service until the late 1940s. During the war, they were leased by *Flygvapnet*.

In the summer of 1940, three Luftwaffe Ju 52/3ms, which had made unscheduled arrivals in neutral Sweden, were civilian registered for the flight to Travemünde, where the aircraft were returned to the Luftwaffe.

With DC-3s replacing the old ABA Ju 52/3ms, four of them were sold abroad in 1949, one to Greece and three to Spain. The only former ABA Ju 52/3m to remain in Sweden was SE-ADR. In 1947, it was sold to the AB Industridiesel company, which transferred ownership of the Ju 52/3m to Albin Ahrenberg the following year. A notable Swedish aviation pioneer, Ahrenberg toured extensively, giving air experience flights in the old Junkers. While his competitors secured higher profits by using more modern aeroplanes, Ahrenberg said that he would 'never abandon an old friend'. Three additional Ju 52/3ms, registered as SE-AYB, -BUD, and -BUE, were acquired from Denmark and Finland. However, the writing was on the wall, and by 1952 Ahrenberg was bankrupt. His Ju 52/3ms were sold to Australian airline entrepreneur Bobby Gibbes (see page 91).

Swedish Civilian Ju 52/3ms

Registration	W Nr	Name	P/i	Date registered	Cancelled	Remarks
SE-ADM	4004		D-2317	13.12.1932, 26.05.1933, 30.08.1933	24.02.1933, 11.06.1933, 25.10.1933	Ju 52 ce. Re-registered D-2317, later D-UBES
SE-ADK(1)	4017	*Södermanland*	SE-ADA, ntu, SE-ADI, ntu	Restored 04.07.1932		Damaged in landing accident 15.07.1932. Repaired and re-registered SE-ADR 15.07.1932. Cancelled 24.08.1955
SE-AER	5440	*Västmanland*		31.03.1936	09.01.1937	Sold to Britain, R G-AERU
SE-AES	5518	*S.A.Andrée*		02.07.1936	02.03.1937	Sold to Britain, R G-AERX
SE-AFA	5614	*Svealand*		12.12.1936	31.08.1949	Sold to Greece, R SX-CDA
SE-AFB	5620	*Götaland*		20.11.1936	18.01.1949	Sold to Spain, R EC-ADO
SE-AFC	5633	*Norrland*		23.12.1936	18.01.1949	Sold to Spain, R EC-ADO
SE-AFD	5646	*Vikingaland*		07.05.1937	18.01.1949	Sold to Spain, R EC-ADP
SE-AKR	6664		Lw SE+IM	13.08.1940	02.09.1940	Int Vallsta 17.04.1940. Ret to Lw 02.09.1940
SE-AKS	5791		Lw SE+IM	13.08.1940	02.09.1940	Int Grums 16.04.1940. Ret to Lw 02.09.1940

SE-AKS was one of three Ju 52/3m interned in Sweden during the spring of 1940. They were issued with temporary Swedish civilian registrations and flown back to Germany on 2 September 1940. (*Arlanda Civil Aviation Collection*)

SE-ADR seen in the early 1950s, when employed in transporting newspapers. Twenty years before, the Ju 52/3m had been the most modern and capable of all European airliners. Interestingly, the passenger seats appear to still be fitted. (*Swedish Aviation Historical Society*)

SE-AKT	6132		Lw SE+HU	13.08.1940	02.09.1940	Int Siverbo 11.04.1940. Ret to Lw 02.09.1940
SE-AYB	641375		Lw, OY-DFU	18.09.1950	30.09.1955	Sold to Australia R VH-BUW
SE-BUD	7493		D-AXVA, OH-LAO, D-AEAC, OH-LAO	26.09.1952	30.09.1955	Sold to Australia. R VH-VH-BUV, later VH-GSS
SE-BUE	7256		D-AVIU, OH-LAM	04.07.1952	30.09.1955	Sold to Australia. R VH-BUU

Switzerland

Interestingly, Switzerland was one of very few European nations that did not to use the Ju 52/3m in regular civilian passenger services, with the national airline Swissair opting to purchase Douglas DC-2s and DC-3s instead. However, the Ju 52/3ms operated by the Flugwaffe were civilian-registered due to some interesting circumstances. Due to the strict neutrality of Switzerland, flights by military aircraft outside of the country were strictly forbidden. Post-war, the trio of Flugwaffe Ju 52/3ms were placed on the civilian register, along with several Messerschmitt Bf 108B *Taifun* and Nord N.1203 Norécrin light-communication aeroplanes. The reason for this was to ferry Flugwaffe pilots and maintenance personnel to Germany and Great Britain in order to collect P-51D Mustangs and DH 100 Vampires bought by the Flugwaffe.

In September 1959, the Ju 52/3ms were all removed from the civilian register. When they were withdrawn from military service in 1982, all three Ju 52/3ms were re-registered, being used for joyrides, movies, and air-show appearances. To this day, the company Ju Air continues to operate Ju 52/3ms, also having added a CASA 352 to its inventory (see Chapter 19 for more details).

Swiss Civilian Ju 52/3ms

Registration	W Nr	P/i	Date registered	Cancelled	Remarks
HB-HOS	6610	A-703	09.05.1947	03.09.1959	Re-registered as HB-HOS on 26.08.1982. Current
HB-HOT	6595	A-702	27.03.1951	03.09.1959	Re-registered as HB-HOT on 15.10.1982. Current
HB-HOP	6580	A-701	13.09.1948	03.09.1959	Re-registered as HB-HOP on 15.10.1982. Current
HB-HOY	96	T.2B-165, D-CIAK		24.07.1997	CASA 352. Current

CASA 352 HB-HOY seen at Mönchengladbach, Germany, on 21 June 2015. (*Author's Collection*)

Turkey

Five Ju 52/3ms were delivered in April 1944 to *Devlet Hava Yollari* (State Air Lines). They were registered as TC-PEK (W Nr 130731), TC-RUH (W Nr 130752), TC-SEL (W Nr 130761), TC-TEZ (W Nr 130862), and TC-ULU (W Nr 130871) respectively. Based at Ankara, DHY operated a mix of de Havilland twin-engined light transports for domestic routes. Following the end of the war, surplus USAAF Douglas C-47s were delivered, which resulted in the Ju 52/3ms being withdrawn from service in 1947.

Uruguay

Formed in 1937, *Compañía Aeronáutica Uruguaya SA* (CAUSA) obtained two Ju 52/3ms the following year. Registered as CX-ABA (W Nr 5877) and -ABB (W Nr 5886), the pair of Ju 52/3ms mostly operated on floats, connecting the capital Montevideo with Buenos Aires twice a day. Apart from passengers and freight, the CAUSA Ju 52/3ms also carried mail, for which the airline received a yearly $U55,000 subsidy during the initial four years of operations. CX-ABA was named *Uruguayo*, and -ABB *El Argentino*, the latters name being changed to *Buenos Aires Herald*. Instead of ordinary passenger seats, benches were fitted that made it possible to carry twenty-eight passengers.

CX-ABB was written-off on Christmas Eve 1940, when it sunk in a gale in Buenos Aires harbour. CX-ABA continued in service, and was reportedly later transferred to the Museo Aeronautico for preservation. Interestingly, the 1948 edition of *Jane's All the World's Aircraft* states that CAUSA was still operating two Ju 52/3ms.

One of the DHY Ju 52/3ms. (*Rob Mulder*)

USA

The first civilian Ju 52/3m to be registered in the USA was the former HC-ABS. A much-travelled aeroplane, the Ju 52/3m had been stored outside since retirement in 1963. Fate intervened when American vintage aviation enthusiast Lester 'Bud' Weaver of Polo, Illinois, began negotiating to purchase the dilapidated Ju 52/3m. Following more than a year of difficult negotiations, Weaver succeeded in his mission in obtaining the ancient transport on 25 May 1970, with the price being $5,000. Restoring the old Ju 52/3m to airworthy condition fell to George Hamilton, the Director of Argosy Airlines. One of the first tasks was allegedly the removal of a Condor's nest from the tail section. For the next five months, Hamilton and Weaver worked hard to get the Ju 52/3m airworthy. The initial test flight took place on 10 November 1970, with the Ju 52/3m having been registered as N130LW. With everything being deemed to be working okay, the ferry flight to the USA began on 14 November. Eight days later, Hamilton and Weaver arrived at Dixon, Illinois, where the Ju 52/3m was inspected by FAA officials. It was on display at the 1972 EAA Oshkosh fly-in. According to a contemporary article 'once it had landed and parked, the curious crowd swarmed about it giving out all kinds of head shakes, bemused grins and remarks of surprised discovery'.[9]

The Ju 52/3m was eventually sold to aviation writer Martin Caidin in 1976. During an overhaul, seven layers of paint were stripped from the fin, revealing a blood-red swastika. It was eventually sold to Air Repair, Opa Locka, FL, from whom Lufthansa bought it in 1984. Completely rebuilt, the Ju 52/ currently flies with the Lufthansa Traditionsflug as related elsewhere.

Two CASA 352s were obtained by the Military Aircraft Restoration Corporation (MARC), being registered as N9012N and N9012P respectively. None of them arrived in America, being traded to European aviation museums.

N52JU seen during the mid-1970s, when owned by Martin Caidin. (*Swedish Aviation Historical Society*)

The Confederate Air Force (CAF) purchased a Spanish-built CASA 352 in 1977. It is currently airworthy with Jerry Yeagen at Virginia Beach, VA.

US Civilian Ju 52/3ms and CASA 352s

Registration	W Nr, C//n	P/i	Date registered	Cancelled	Remarks
N130LW, N52JU	5489/130714	D-AQUI, LN-DAH, Lw, D-AQUI, LN-KAF, HC-ABS	11.1970		Preserved in Germany
N9012N	33	T.2B-142	24.10.1976	17.05.1993	Preserved in Sweden
N9012P	31	T.2B-140	26.04.1984	11.08.1992	Preserved in Germany
N352JU	67	T.2B-176			Initially registered N99069. Preserved in USA

USSR

Civilian use of the Ju 52/3m in the USSR was quite extensive, lasting from the mid-1930s until the early 1950s. It was, however, divided into three distinctly separate usages. The first of these was the Soviet-German airline Deruluft, which had five Ju 52/3ms in service from 1934 to 1936. When Estonia was occupied in 1940, the pair of civilian Ju 52/3ms operated by the airline AGO was taken over by Aeroflot. The third use occurred in 1941, when a few Ju 52/3ms were ordered from Junkers for use as engine test beds. During the war, large quantities of Ju 52/3ms were captured and transferred to civilian use. The remaining civilian Ju 52/3ms were struck off charge in the early 1950s.[10]

Following the signing on 6 May 1921 of a provisional trade agreement, concession was given for the establishment of a German-Soviet airline. The new airline, named Deutsch-Russische Luftverkehrsgesellschaft (German-Russian Air Traffic Company), abbreviated to Deruluft, received a concession on 17 December 1921 to operate the Moscow–Königsberg route. In 1925, the route was extended to Berlin, with the Königsberg–Berlin leg being flown from the following year in cooperation with the newly established *Deutsche Luft Hansa AG*.

In all, five Ju 52/3ms were operated by Deruluft, and the transports retained their German civilian registrations. D-AREN arrived in 1934, but was written-off on 31 January 1935 near Stettin. Four additional Ju 52/3ms were subsequently transferred to Deruluft from Lufthansa stocks in 1935. At the time, Deruluft also operated three Junkers F.13s, three Rohrbach Rolands (one being lost in March 1935), and four Tupolev ANT-9s. Deruluft's concession expired on 31 December 1936, which saw the return to Lufthansa of the Ju 52/2ms. The end of Deruluft was due to several different reasons—a reduced demand for civilian air traffic between Germany and the USSR, and stricter security measures from both countries. Nevertheless, six chartered aircraft, including three Ju 52s, continued to operate Deruluft's routes until 31 March 1937. In 1935 and 1936, Deruluft carried 32,879 passengers, 176,110 kg of mail, and 862,130 kg of other cargo and baggage.[11]

Deruluft Ju 52/3ms

Registration	W Nr	Name	ToC	SoC	Remarks
D-AREN	4051	*Phönix*	1934	1935	Deruluft W/o Stettin 31Jan 35. 11 killed
D-AXES	4052	*Kondor*	02.1935	31.12.1936	Deruluft, returned to DLH
D-AHUS	4049	*Milan*	03.1935	31.12.1936	Deruluft, returned to DLH
D-AGIS	4048	*Kormoran*	04.1935	31.12.1936	Deruluft, returned to DLH
D-ADAL	4046	*Flamingo*	05.1935	31.12.1936	Deruluft, returned to DLH

Following the incorporation of the Baltic States in 1940, the pair of Ju 52/3ms of the Estonian airline AGO were transferred to Aeroflot. They were handed over to the Baltic Department in September 1940, and subsequently registered as CCCP-Л22 and CCCP-Л23 in February 1941. Following the German invasion of the USSR, the pair of Ju 52/3ms were operated on the Moscow–Irkutsk route. In late 1942, a few captured former Luftwaffe Ju 52/3ms were received by Aeroflot. One necessary requirement for operating the Ju 52/3ms was that they would be used far away from the front in order to avoid misidentification with Luftwaffe aircraft. With many Ju 52/3ms being captured, three repair depots as well as factory No. 243 at Tashkent were involved in the rebuild and maintenance work of Ju 52/3ms. Although many of the aircraft could be restored to airworthiness, others were only suited for spares recovery. Some parts, such as radios, carburettors, tires, etc., were replaced by standard Soviet equipment. One Ju 52/3m was even re-engined with M-621R engines. In 1943, thirty-one Ju 52/3ms were transferred to Aeroflot, with known registrations being CCCP-Л26, -Л27 (W Nr 7341), -Л28, -Л30 (W Nr 801250), -Л31 (W Nr 808999), -Л32 (W Nr 2901), -Л33 (W Nr

7586), -Л34 (W Nr 809899), -Л35 (W Nr 6445), -Л36, -Л37 (W Nr 6486), -Л38, -Л39 (W Nr 2897), -Л40 (W Nr 3325), -Л41, -Л42 (W Nr 05210021), -Л43 (W Nr 5192), -Л44, -Л45 (W Nr 7189), -Л46 (W Nr 6053), -Л47, -48 (W Nr 7598), -Л49 (W Nr 3092), -Л50, -Л51 (W Nr 6538), -Л52, -Л53, -Л54 (W Nr 7335), -Л55 (W Nr 7082), -Л56 (W Nr 3244), -Л57 (W Nr 7512), -Л58 (W Nr '58'), -Л59 (W Nr 7568), -Л60 (W Nr 4069), -Л61 (W Nr 5457), -Л61 (second use, W Nr 5882) -Л62 (W Nr 5338), -Л63 (W Nr 6717), -Л64, -Л65, -Л66, -Л67, and -Л68. As described in the survivor's section, CCCP-Л54 was recovered from its Siberian crash site in 2003.

Routes flown included Sverdlovsk–Krasnoyarsk and Kuybyshev–Tashkent–Almaty. In late 1944, all remaining Ju 52/3ms were transferred to the Glavnoe Upravlenie Grazhdanskogo Vozdushnogo Flota (Chief Directorate of the Air Fleet, GUGVF). Following the German surrender, thirty-seven Ju 52/3ms and no less than 137 BMW 132 engines were delivered. An additional five Ju 52/3ms had been confiscated in Romania, with one more following in 1946. Large quantities of spares meant that further Ju 52/3ms were taken into service. Most of the Ju 52/3ms were used as pure freighters. Operating in primitive conditions meant a fair amount of wear and tear, with the final Aeroflot Ju 52/3ms being struck off charge in 1949. A few other government operators of the venerable Ju 52/3m were Polyarnaya Aviatsiya (Polar Aviation), the Ministry for Internal Affairs, and the Ministry for Fishery Industry. All of these had been replaced by the early 1950s.

On 18 February 1941, ten Ju 52/3ms were ordered. The intention was to utilise the Ju 52/3ms as engine test beds. Four were delivered in March–April (including W Nr 7120, 7121, and 7180). Among the operators of these Ju 52s was factory No. 26 at Rybinsk, which also received two former Luftwaffe Ju 52/3ms.

Yemen

One AAC.1 (c/n 73) was purchased on 13 April 1948 by representatives of the Yemeni King, Ahmad al-Shams, from the Lebanese flag-carrier Air Liban. It was assigned the registration YEMEN 1. It would appear, however, that the aircraft did not reach Yemen. One possibility is that the AAC.1 ended up in Syria.

Yugoslavia

Originally formed in 1927, the airline Aeroput ceased operations in April 1941. Following the end of the war, JAT Airways—*Jugoslovenski Aerotransport*—began operations on 1 April 1947. Based at Zemun Airport, Belgrade, JAT operated three Ju 52/3ms/AAC.1s. The only aircraft identified is AAC.1 YU-ACE (c/n 316), which crashed on 29 June 1951 soon after take-off from Rijeka—all fourteen aboard were killed. The other JAT aircraft may have been AAC.1s c/n 77 and 78, which were delivered to Yugoslavia on 15 August 1945.

18

Military Operators

Between 1932 and 1982, a large number of air arms operated the venerable Ju 52/3m. The largest military operator was, of course, the German Luftwaffe (described separately). First seeing action in Bolivia during the Chaco War, the Ju 52/3m subsequently served in Europe and North Africa. Interestingly, apart from the Luftwaffe, Bulgarian, Hungarian, Italian, and Romanian air arms, four Allied air forces counted Ju 52/3ms among its inventory. The first of these were the South African Air Force (SAAF), which used both aircraft requisitioned from South African Airways as well as captured Luftwaffe aircraft. The RAF operated a few captured Luftwaffe aircraft in 1943–44, with more than 100 being allocated RAF serials following the end of the war. Few of these surrendered aircraft saw RAF use, instead being held in reserve for an anticipated requirement from various British airlines. The USAAF operated one solitary Ju 52/3m, designated as the Junkers C-79. Finally, the French Air Force (*Armée de l'Air*) received one or more captured Ju 52/3ms in 1943. Following liberation of Paris, the Ju 52/3m production line was kept open with many aircraft being supplied to the *Armée de l'Air* and the *Aéronautique navale*, seeing extensive use in Indochina and Algeria. Interestingly, a few French aircraft saw brief service during the Berlin Airlift.

Portugal, Spain, Sweden, and Switzerland, all neutral and unaligned during the war, flew Ju 52/3ms, with the latter not relinquishing their aircraft until the early 1980s. As mentioned above, Ju 52/3ms were operated by the Bolivian air arm, as well as in Argentina, Colombia, Ecuador, and Peru.

Argentina

In 1935, the Argentinian Air Force and Navy evaluated the Ju 52/3m bomber. However, the American Martin 139 was selected as a standard bomber. In the event, a number of Ju 52/3ms were nevertheless to serve with the Servicio de Aviación Militar (Argentine Army Air Force, SAM). Five were ordered from Junkers, all being delivered in September 1939. Initially, the Ju 52/3ms serviced with the *Agrupación Instruc. De la Escuela Application Ejercito*. In 1941–42, the aircraft were transferred to Agrupación Transporte, eventually serving with *Línea Aéreas del Sodoeste* (South-West Airlines, LASO), a military run airline. LASO was eventually superseded on 23 October 1945 by *Líneas Aéreas del Estado* (State Airlines, LADE). On 4 January 1945, an independent air arm, the Argentine Air Force (*Fuerza Aérea Argentina* (FAA)) was formed.

In 1946–47, the original Ju 52/3ms were augmented with eight additional aircraft, purchased from the Brazilian airline Syndicato Condor. Most were used by LADE, eventually being transferred to the *Ministerio de Agricultura y Ganaderia* (Ministry of Agriculture and Livestock Raising) in early

FAA Ju 52/3m s/n 164 being refuelled. (*Michael Magnusson*)

1949 for use as crop dusters. In this context, T-166 had been converted as a crop duster already in 1947. By this time, the original BMW engines were becoming difficult to maintain, which resulted in some aircraft being fitted with Pratt & Whitney R-1340 engines.

Two Ju 52/3ms were retained by FAA (with two more being returned from the Ministerio de Agricultura y Ganaderia in 1954). T-153 was subsequently transferred to the aircraft manufacturer Instituto Aerotécnico (I.Ae.) for use as a glider tug. On 1 October 1954, the I.Ae. 37P glider was towed into the air (designed by Reimar Horten, the intention was to develop the I.Ae. 37P into a jet-powered research aeroplane, capable of reaching Mach 1.2. This was eventually cancelled due to lack of funds). Apart from glider towing, T-153 was also used as an engine test bed for the I.Ae. 19 El Indio piston engine. The Ju 52/3m was eventually lost during a test flight on 12 October 1957. The second Ju 52/3m, T-159 suffered a mid-air collision with an FAA DH 104 Dove 8 km south of Villa Mugueta on 12 November 1953, resulting in twelve deaths.

Argentinian Military Ju 52/3ms

Serial	W Nr	Delivered	Name	Remarks
164, T-164	6264	16.09.1939	*Nahuel Pan*	R LV.YSI 29.12.1950
165, T-165	?	16.09.1939	*Coronel Fontana*	Wfu 22.11.1946
166, T-166	6224	16.09.1939	*Neuquen*	R LV-ZBL 08.02.1949

167, T-167	6403	25.09.1939	*Cataratas del Iguazú*	R LV-ZBF 08.02.1949. Ret to FAA 04.08.1954
168, T-168	6354	25.09.1939	*Gral San Martin*	P/I D-AGCB. R LV-ZBJ 08.02.1949
T-152	5053	03.02.1946	*Santa Cruz* ?	P/i D-AQUQ, HC-SAC, PP-CBR. R LV-ZBG 08.02.1949
T-153	4078	04.01.1946	*Misiones*	P/i D-AKIQ, PP-CBB. To FMA as glider tug and engine test bed. W/o near Cafayate 12.10.1957
T-154	5478	03.02.1946		P/i D-AJAO, PP-CBD. Wfu 27.02.1947
T-155	4079	03.02.1946	*La Pampa*	P/i D-APOR, PP-CBF. R LV-ZBH 08.02.1949. Ret FAA 54
T-156	4075	03.02.1946	*Neuquen*	P/i D-APEF, PP-CBG. R LV-ZBI 08.02.1949
T-157 *	5120	08.08.1947		P/i D-ADER, PP-CBE. R LV-ZBM 08.02.1949. NB; possibly serial T-169
T-158	4043	08.01.1946	*Yapeyu*	P/i D-3356, D-ABIS, PP-CAX. R LV-ZBD 08.02.1949. Preserved
T-159	5261	08.01.1946	*Rio Negro*	P/i D-AGST, PP-CAZ. Mid-air collision with DH 104 Dove T-82 12.11.1953, Crashed Villa Mugueta.
T-160	4038	08.01.1946		P/i D-ARUW, PP-CAV. R LV-YSH 29.12.1950
T-161	5656	08.01.1946		R LV-ZBE 08.02.1949
T-169 *	5120	08.08.1947		P/i D-ADER, PP-CBE. R LV-ZBM 08.02.1949. NB; possibly serial T-157

NB: The 'T' prefix was added on 22.07.1944

Austria

The interwar Austrian Air Force (*Österreichische Luftstreitkräfte*) purchased three Ju 52/3ms. Two of these were bombers with the third Ju 52/3m being assigned for VIP transport duties. All three were delivered in July 1937. The VIP transport Ju 52/3m (W Nr 5835) was allocated the military serial number 32 as well as the civilian registration OE-HKA. The pair of bombers (W Nrs 5807 and 5820) received the serials 36 and 37. Following the Anschluss, i.e. the incorporation of Austria into Germany on 13 March 1938, the Ju 52/3ms were transferred to Luftwaffe.

Bolivia

The first use of the Ju 52/3m in a war did not occur in Europe, but in Bolivia. Since 1928, Bolivia and Paraguay had waged a bloody war over a contested border area. The war, known as the Chaco War, saw the application of air power, which included airlift operations. In the summer of 1932, Bolivian forces initiated a large-scale offensive. As described in Chapter 17, Lloyd Aero Boliviano

(LAB) had obtained several Ju 52/3ms. The first Ju 52/3m arrived in Bolivia in November 1932 with two more following two months later. The three LAB Ju 52/3ms were requisitioned by the Cuerpo de Aviación to deliver supplies to isolated garrisons, as well as evacuating wounded soldiers. At least two of the Ju 52s, named *Chorolque* and *Huanuni*, were camouflaged with green and brown blotches, with the Bolivian flag appearing on the rudder—the name of the aircraft was painted on the fuselage. Until the end of the war in 1936, the Ju 52/3ms airlifted some 40,000 troops wounded in battle and 4,400 tons of cargo. Other missions included transporting the Bolivian President Salamanca (in *Chorolque*) to inspect troops in the Chaco region. In Bolivia, the hardworking Ju 52/3ms were colloquially known as *Los Trimotores*.

Apart from the LAB trimotors, one Ju 52/3m was donated to the *Cuerpo de Aviación* by Simón Patiño, a wealthy tin ore magnate. This particular Ju 52/3m (named *Bolivar*) was subsequently operated as a VIP transport for the German-born Commander of the Bolivian armed forces, Hans von Kundt. One interesting mission occurred when *Bolivar* was used to fly a delegation to the Paraguayan capitol Asuncion to mediate in the conflict. When the Chaco War ended, the Ju 52/3ms returned to carrying civilian passengers and cargo.

Bulgaria

Two Ju 52/3m g3es were delivered in 1939 to Royal Bulgarian Air Force (*Vazhdushnite na Negovo Velichestvo Voiski* (VNVV)). In January 1940, the pair of Ju 52s, registered as LZ-UNA and -UNB, were attached to the Instrument Flying School, based at Plovdiv. An additional two Ju 52/3m g10es were delivered in 1943, with four being on charge in September 1944 with the Prenosno Yato (Air Transport Squadron). The month before, Bulgaria had switched sides and declared war on Germany. As a result, Bulgarian forces conducted operations in the Balkans and Yugoslavia. In September, VNVV Ju 52s flew thirteen sorties over Ratunda-Drenak, in present-day Macedonia, air-dropping about 14 tons of supplies to the encircled Bulgarian forces. Two of the Ju 52s were subsequently attached to a liaison and transport squadron, based at Vrazhdebna, for support of the headquarters of the field armies. In the first months of 1945, the Ju 52s were used in support of the Bulgarian 1st Army operations in western Yugoslavia. During the summer of 1945, several former Luftwaffe Ju 52s were found at Zeltweg, Austria, and flown to Bulgaria.

On 20 April 1946, one Ju 52/3m transported a Bulgarian delegation to Copenhagen, Denmark. With Bulgaria quickly becoming an ardent ally of Stalin's USSR, the VNVV was renamed the Vazdusni Voiski (VV). On 15 September 1946, the Bulgarian People's Republic was declared. In September 1949, the name of the Bulgarian air arm was amended to Voennovazdushni Sili.

On 19 March 1947, three Ju 52s were transferred to the recently established Direktsiya za Vŭzdushni Sŭobshteniya (Air Communications Directorate DVS). Commanded by Colonel Mikhalakiev (former C-in-C of the People's Air School), the DVS remained under direct government control, with one of the Ju 52s registered as LZ-UNL. On 30 June 1948, Mikhalakiev hijacked a DVS Ju 52/3m, killing the pilot General Ganev and injuring other members of the crew. Mikhalakiev then flew the Ju 52/3m to Istanbul, where he was granted political asylum.

In December 1949, six Ju 52/3ms remained in VV service, five with the 14th Air Transport Polk, based at Vrazhdebna, and one with the 3rd Advanced Training Escadrila of the People's Air School, based at Telish. In 1950, the remaining TABSO Ju 52/3ms were returned to the VV inventory. The venerable Ju 52/3ms were used by the air experience and auxiliary flying training unit. The final Ju

52/3m were eventually withdrawn from service during the early 1950s, being replaced by Lisunov Li-2 Cabs (i.e. Russian-built Douglas DC-3s).

China

In 1931-32, the Junkers K 45 bomber, as well as the Ju 52/3m transport, was offered to the Nanking government—this offer was rejected. In 1935, another offer for six float-equipped K 45Ws was also declined. One Ju 52/3m was used by the Chinese Nationalist Air Force (CNAF) as a VIP transport. The aircraft in question was an early production example, W Nr 4044, originally operated by Lufthansa as D-ABAN. The German Ministry of Propaganda decided to despatch one Ju 2/3m to China to undertake a lengthy tour of the country. Taking off from Berlin on 29 August 1934, it arrived in Shanghai eight days later, on 6 September 1934. The Ju 52/3m was subsequently sold to the Chinese government in April 1935 for the specific use of VIP transport. The Ju 52/3m was used on several occasions by Chiang Kai-shek. Its ultimate fate is unknown.

Colombia

The second-largest military operator of the Ju 52/3m in South America was the Colombian air arm, *Aviación Militar*, who purchased Six Ju 52/3m bomber transports.

When, in 1932, a military conflict erupted with neighbouring Peru over the contested border town of Leticia, both countries scrambled to buy warplanes. Six Ju 52/3ms were ordered for the *Aviación Militar* through the auspices of the airline SCADTA (Sociedad Colombo Alemana de Transportes Aéreos, Colombian-German Air Transport Company). SCADTA had been established in late 1919, and had, since then, mainly operated aircraft of German origin. Among its founders were resident German businessmen. The chief pilot of SCADTA, Herbert Boy (a former fighter pilot of Jasta 14), was ordered by the Colombian president to establish an aerial supply route to Leticia and its surrounding areas along the Putumayo River. In order to achieve this, additional, larger aircraft were needed. The Ju 52/3m fitted the bill perfectly, being able to operate on floats or wheels, and also having the capability of carrying a large amount of cargo and to function as a bomber. Four Ju 52/3ms soon arrived, being assigned the serials 621 to 624. No bombing sorties were flown, however, although a number of supply flights took place. Initially, the Ju 52/3ms were based at Palanquero.

The delivery of the last two of these aircraft was delayed, reportedly due to the intervention of Hermann Göring, who considered the then still-secret Luftwaffe to have priority. Major Herman Boy of the *Aviación Militar* subsequently arrived in Berlin, where he persuaded his former class mate Joseph Goebbels to assist in the release of the trimotors. By the summer of 1933, through the mediation of the League of Nations, the conflict ended. On 25 June, the town of Leticia was formally handed over to Peru.

The final two Ju 52/3ms did not arrive until March 1934, receiving the serials 625 and 626 respectively. Two were written-off in crashes in 1936, serial 623 at Cabuyaro and 624 at Putumayo. Mostly operating on floats, the four surviving Ju 52/3ms, serials 621, 622, 625, and 626, were still in service in February 1943. The final Ju 52/3ms lasted until 1950, when they were withdrawn—one has been preserved.

Croatia

One Ju 52/3m was presented to Croatian dictator Ante Pavelić as a VIP transport. On 28 December 1944, the Ju 52/3m was destroyed at Borongaj when partisan forces attacked the airfield.

Czechoslovakia

The Czezh Air Force (*Československé Vojenské Letectvo* (CVL)) operated twelve (or, according to one source, thirteen) Ju 52/3ms from 1945 until the early 1950s. On 26 May 1945, the *Letecká Dopravní Skupina* (Air Transport Group, LDS) was formed at Prague Ruzyně to secure national and international air transport requirements for the Czech government. Initially, five Ju 52/3ms, refurbished by aircraft manufacturer LETOV, were used (all CVL Ju 52/3ms were refurbished by LETOV). In CVL service, the Ju 52/3m was designated D-7, 'D' indicating *Dopravní letadla*, i.e. transport aircraft.

In early 1946, three Ju 52/3ms were transferred to the national airline CSA, with the LDS being renamed on 1 April as the *Letecký Dopravní Pluk* (Air Transport Regiment 1, LDP). On 15 January 1947, the LDP 1 had seven Ju 52/3ms in service. The Ju 52/3ms were used quite extensively, with 1,185 flight hours being accumulated in 1947 during 2,643 flights. In May 1947, the CVL designation was changed from D-7 to D-52. Known individual serial numbers were D-1, D-2, D-3, D-4, D-5, D-7, D-8, D-9, D-10, D-11, D-12, and D-19. On 15 March 1948, eight Ju 52/3ms (or rather D-52s) were in use, rising to nine on 1 February 1949. Some three months later, on 15 May, the LDP was renamed as the LDP 1, indicating the formation of a second air transport regiment. The Ju 52/3ms were gradually withdrawn from use, with four remaining in service on 1 November 1950. In 1951, five Ju 52/3ms, with the serials D-20, D-40, D-41, D-42, and D-44, remained in use, all being withdrawn from use shortly afterwards.

One Ju 352 also entered service with the CVL, receiving the serial D-352. It was later civilian registered as OK-JUE, and used on a Prague–Moscow flight. Aboard the Ju 352 was the first Tatra car produced after the end of the war. The Tatra was presented as a gift to Stalin. The Ju 352 returned to Czechoslovakia, but its final fate remains undetermined.

One solitary Ju 352 was briefly used by the Czechoslovakian Air Force. (*Miro Patocka*)

East Germany

In September 1952, initial flight training of 271 pilots belonging to the *Volkspolizei-Luft* (VP-Luft) began in the USSR. Subsequent training took place at Cottbus, Bautzen, and Kamenz. The VP-Luft pilots were subordinated to the *Kasernierte Volkspolizei* (KVP). All VP-Luft aircraft, including Yakovlev Yak-18 and Yak-11 trainers, carried Soviet Red Star insignia. At least one Ju 52/3m was operated by the VP-Luft. On 18 March 1953, a Ju 52/3m crashed soon after take-off from Cottbus-Drewitz Airport. The Junkers was burnt out, and all eleven aboard were killed. It is certain that no Ju 52/3ms were operated by the *Nationalen Volksarmee* (National People's Army, NVA), which was formed in March 1956 as the successor of VP-Luft.

Ecuador

Two Ju 52/3ms of SEDTA, HC-SAD (W Nr 5109) and -SAE (W Nr 5283), were requisitioned by the Ecuadorian government on 5 September 1941. Both were transferred to the *Fuerza Aérea Ecuatoriana* (FAE), where military markings were applied—the names *Azuay* (HC-SAD) and *Aconcagua* (HC-SAE) were retained. The Ju 52/3ms were the largest aeroplanes operated by the FAE. With the commencement of US Lend-Lease deliveries and agreement for the establishment of US air bases and other military installations on Ecuadorian territory, one of the Ju 52/3ms, *Aconcagua*, was formally transferred to the USAAF in May 1942. The second Ju 52/3m was retained by the FAE, remaining in service until the early 1950s, when it was disposed of to the local airline TAO.

Finland

Following the outbreak of war on 30 November 1939, Finnish forces fought valiantly against superior Soviet forces. With the Finnish Air Force, *Ilmavoimat*, having few dedicated transport aircraft, the two Ju 52/3ms of Aero O/Y were leased, although not simultaneously.

On 1 December 1939, OH-LAK (W Nr 4014) was taken over. Having flown 118 hours, the Ju 52/3m was handed back to Aero O/Y on 21 February 1940. It was replaced by the second Aero O/Y Ju 52/3m, OH-ALL (W Nr 5494), which was leased by *Ilmavoimat* between 21 February and 19 March 1940. OH-ALL was quite heavily used, accumulating 134 hours before being returned to Aero O/Y. Both Ju 52/3ms were assigned to T-LeR 4 (Aviation Regiment 4). The Winter War ended on 13 March 1940 with much border territory being handed over to the USSR.

During the Continuation War, the Aero O/Y fleet was placed under military control, being named Civil Aviation Fleet of the Air Forces. However, the aircraft retained their civilian registrations.

France

The Junkers Ju 52/3m, including the French-built AAC.1, was to remain in French military service for nearly twenty years, far longer, in fact, than the Luftwaffe. The first Ju 52/3m to enter service was one (or more) example captured in Tunisia in 1943 by Free French forces, remaining in service until 1945 at least. Following the end of the war, the *Armée de l'Air* (French Air Force) received 417 former Luftwaffe aircraft from the RAF, including thirty-six Ju 52/3ms. Nine of the Ju 52/3ms were fitted with floats. The aircraft were refurbished at *Armée de l'Air* depots at Leck and Lüneburg. Those not returned to flying condition were dismantled and sent to France by rail. The responsible *Armée de l'Air* unit was the *Détachement de Désarmement Aérien*, which was subordinated to the *Service des Affaires Allemandes*. Subsequently, thirty-eight Ju 52/3ms were taken on charge, with their respective *Werk Nummer* serving as serial numbers. These Ju 52/3ms subsequently underwent refurbishment, and were issued with new construction numbers—1001 to 1038. All of them were delivered to the French military. One reason for the discrepancy between this quantity and the number supplied from RAF stocks could be that two Ju 52/3ms were already in French service. Additionally, seven Ju 52/3ms operated by *Armée de l'Air* used their respective *Werk Nummern* as identities—W Nr 0645, 0703, 1109, 3376 (RAF serial VM928), 3412 (RAF serial VM932), 313558, and 501367 (RAF serial VM918).

With large numbers of Ju 52/3ms having been built in France during the war, the existence of jigs, spares, etc., saw the continuation of production of the German transport until 1948. The company responsible was Ateliers Aéronautiques de Colombes (AAC), with the French-built Ju 52/3m being designed AAC.1, using BMW 132Z engines. In all, no less than 415 AAC.1s were completed, with the majority being delivered to the *Armée de l'Air* and *Aéronautique navale*. The *Armée de l'Air* called the new model the AAC.1 Toucan, with a number of *Groupes de Transport* (GT), including GT 2/61 *Maine*, GT 1/62 *Algerie*, GT 2/62 *France-Comte*, GT 1/63 *Bretagne*, GT 1/64 *Bearn*, and GT 3/64 *Tonkin*, being equipped with the type. In more ways than one, the Toucan was the true workhorse of *Armée de l'Air* during the late 1940s and early 1950s, along with the Douglas C-47 Skytrain, upon which the development of military transport aviation was based after the end of the Second World War. The AAC.1 was of rugged construction and easy to maintain, with the *Armée de l'Air* and Aéronavale using their AAC.1s for essentially the same tasks as the Luftwaffe. During the 1950s, the AAC.1 wheel hubs were modified to accommodate Douglas DC-3 tyres. The final *Armée de l'Air* Toucan in service, c/n 86, was struck off charge on 21 March 1962.

Only weeks after the end of war in Europe, on 21 May, the French Air Ministry centralised all military air transport assets to be used for the repatriation of French prisoners of war and civilians that were deported as forced labour for the German war industry or to concentration camps. Repatriation of civil servants and their families that had spent the war in the French African and Asian colonies was to be performed as well. As a result, the French air transport squadrons were brought together within the *Groupement des Moyens Militaires de Transport Aérien* (GMMTA). It was initially intended to disband the GMMTA, in essence an *ad hoc* Air Transport Command, following the conclusion of the repatriation flight. In the event, the GMMTA was not disbanded as a result of the resurgence of national liberation movements and rebellions, particularly in French Indochina and North Africa. A *Sous/Groupement des Moyens Militaires de Transport Aériens* (SGMMTA) was formed, particularly for operations in North Africa and Indochina (incidentally, the GMMTA did not disband until 1961, being replaced by the *Commandement du Transport Aérien Militarie* (COTAM).

By mid-1945, GT 2/15, with its Toucans as well as other air transport squadrons, was involved in repatriating French nationals from Germany (including the French, British, US, and Soviet sectors) and other countries, including Austria, Czechoslovakia, Italy, Norway, Poland, Romania,

Sweden (some 1,700 French nationals were cared for by the Swedish Red Cross), and Yugoslavia. By October 1945, 16,000 people had been brought back to France.

Two additional squadrons, GT 4/15 (previously flying Beech C-45s) and GT 1/62, formed at Algiers in November 1945, operated Toucans. Apart from the repatriating flights, many transport flights that could not be handled by regular civilian airlines were made. This mainly involved flights between Metropolitan France and its colonies in North Africa, with GT 1/62 in particular being heavily involved on such operations.

On 1 July 1947, GMMTA was reorganised, resulting in GT/15 being renumbered GT 2/61 *Maine* and GT 4/15 becoming GT 3/61 *Poitou*. GT 2/61 was based at Le Bourget and GT 3/61 at Chartres. Based at Biên Hòa in French Indochina, GT 1/34 became GT 1/64 *Béarn*.

During the Berlin airlift of 1948–49, the *Armée de l'Air* made a small contribution. With *Armée de l'Air* transport squadrons being heavily committed to the war in Indochina, only three Toucans of GT 2/62 Maine were available. A number of airlift sorties were made in December 1948. With the decision to standardise on four-engined aircraft, such as the Douglas C-54 and Avro York, the Toucans were withdrawn. However, a more significant French contribution was the expansion of Tegel Airport. Led by French engineers, a workforce consisting mostly of German women expanded the Tegel runway to a length of 2,428 metres (the longest in Europe at the time) as well as the airport's infrastructure, with the work being completed in less than three months.

Toucan Operations in Madagascar

Following the outbreak of an insurrection in Madagascar in March 1947, AAC.1s were used in both the transport and bombing role, with a number of paratroop drops also made. Initially, only five battalions of troops were based in Madagascar, supported by three Toucans and a number of liaison aeroplanes. Reinforcements were immediately despatched from Algeria, with paratroopers being flown in Halifax transports via El Adem, Khartoum, Nairobi, and Dar es Salaam. With the runway at Ivato, Madagascar, being too short to accommodate the Halifaxes, Douglas C-47s were used for the final leg. Air support for the French forces initially consisted of five Avro Ansons of EOM 84, which was normally based at Pointe Noir, French Congo. This quickly proved insufficient, with a squadron of AAC.1 Toucans being formed with aircraft transferred from GT 2/61 and GT 3/61. Having arrived in Madagascar after crossing the African continent, the Toucans were used for aerial reconnaissance, leaflet dropping, medevac flights, as well as airborne assault sorties.

The first airborne assault occurred on 24 April 1947, when paratroopers were dropped over Vatomandry. With the rebels encircling Nosy Varika, paratroopers were dropped from a height of 180 metres, with five paratroopers exiting the Toucans during each low pass. Following this, the Toucans landed at a small football field to airlift women and children out of Nosy Varika. Another paratroop sortie took place at Toamasina on 15 May. A tropical storm on 30 May hindered paratroop drops over Andevarente. Troop reinforcements were brought in by Toucans to Ambalavao on 10 and 11 June.

On 1 July, French forces began a search operation along the north-eastern coast of the huge island. A planned parachute drop against the rebel headquarters at Mangoro had to be postponed due to very poor weather. A parachute drop on 7 July against Anosibe, using six Toucans, went badly wrong when the paratroopers landed in a swamp.

The Toucans were also pressed into action as makeshift bombers. Compared with the Toucans already operating in the bomber role in Indochina, a different method of bombing had to be utilised. With no underwing bomb shackles available, a Toucan bomb sortie involved having 1 ton of 90 kg bombs carried inside the fuselage. Once over the target, the pilot would fly at an altitude of 400 metres and a speed of 120 kph. When seeing the target slip behind the nose of the aircraft, the pilot would bank 5 degrees to the left, count to three, and then order the loadmaster to roll the bombs out of the aircraft. If 50-kg anti-personnel fragmentation bombs were used, the bombs were dropped from 700 metres. Villages suspected of supporting the rebels were bombed at random.

By the end of 1947, most Toucans had been returned to France. Only a few aircraft of GT 2/61 were retained in support of 'mopping-up' operations. By March 1948, the anti-French rebellion was over. The Toucans had operated with a minimum of technical support, with aircraft regularly being returned to Algeria or France for maintenance.

Indochina

Following the Japanese surrender, France was quick to return to its colonies in South East Asia. By autumn of 1945, French colonial administration had been restored to Indochina, i.e. Cambodia, Laos, and Vietnam. However, since 1940, things had changed. A Vietnamese liberation movement, led by Hồ Chí Minh, had declared Vietnam's independence on 2 September. A guerrilla war between Việt Minh and French forces would soon erupt, lasting until 1954 and resulting in the defeat and withdrawal of French forces from Indochina.

The first *Armée de l'Air* units to arrive in Indochina were two fighter squadrons, GC I/7 and II/7. With pilots and ground crews arriving ahead of the aircraft, surrendered Japanese aeroplanes, including Nakajima Ki-43 *Oscar* fighters were pressed into service. A few transports, including several Tachikawa Ki-54 *Hickory* and one Showa L2D3 *Tabby* (licence-built Douglas DC-3) saw brief use, but were soon withdrawn due to their generally poor condition and lack of spares.

In February 1946, the GT I/34 air transport unit arrived, having a strength of sixteen AAC.1s. Based at Biên Hòa, GT I/34 established the first military air links between French strongholds. GT I/34 would be followed by several more air transport units, most of which were equipped with Toucans. These included GT 2/61 *Franche-Comté*, GT 2/15 (eventually renamed *Groupe de marche en Extrême-Orient*, later becoming GT 2/64), GT 2/62 *Franche-Comté*, based at Haiphong and Hanoi, GT 1/64 *Bearn*, based at Nha Trang and GT 2/63 *Senegal*, based at Nha Trang.

Apart from in-country duties, each transport squadron was required to maintain a twice-weekly flight between France and Indochina, known as Line 102, with the route being Paris–Istres–Tunis–El Adem–Cairo–Tehran–Karachi–New Delhi–Rangoon (now Yangon)–Saigon. With the C-47s usually making such a long distance flight in six days, the Toucans with their lower speed and shorter range needed nine or ten days.

On 20 November 1946, French and Việt Minh forces clashed over the control of the customs building at Haiphong. Fighting spread throughout northern Vietnam, with the Việt Minh leader Hồ Chí Minh calling for nationwide resistance against the French exactly one month later, on 20 December 1946, with large-scale attacks against French troop concentrations in northern Vietnam took place. Along with other aircraft, AAC.1s were used to resupply the French garrison in Hanoi. Toucans flew some 150 sorties, landing at Gia Lâm and Bạch Mai, delivering munitions and supplies, as well as evacuating wounded soldiers. Indeed, General Morliére later said that 'Hanoi had been saved by the Ju 52s'.

Although the primary role of the AAC.1 was as a transport, it was also used as a level bomber. Makeshift bomb racks were fitted underneath the fuselage and wings. A maximum of twelve 50-kg

An AAC.1 dropping napalm over suspected Việt Minh strongholds. (*Albert Grandolini*)

bombs could carried (as opposed to early Luftwaffe and *Legion Condor* Ju 52/3ms, *Armée de l'Air* Toucans could not carry bombs internally). Operations against Việt Minh forces included regular bombing sorties. Apart from high-explosive bombs, napalm canisters were also regularly dropped against Vietnamese villages suspected of harbouring Việt Minh guerrillas. The napalm canisters were locally manufactured, and carried inside the fuselage. During operational sorties, they were pushed outside through the open cargo door, a rather imprecise method. No bomb sights other than the Mark I Eyeball were available. Such raids did nothing to enhance the local population's view of the French presence in Vietnam. With Việt Minh having few anti-aircraft guns, *Armée de l'Air* aircraft losses were initially light.

On 5 January 1947, Operation Dedale was initiated. This involved dropping paratroopers over Nam Định, under siege by 5,000 Việt Minh irregulars. The first wave consisted of ten C-47s and ten Toucans dropping paratroopers. C-47s and Toucans subsequently flew 561 sorties in support of the French troops. During forty-one sorties, the Toucans were hit by ground fire on thirty occasions. In the event, French forces progressively regained control of the border area along the Vietnamese-Chinese border. On 9 October, a paratrooper assault against Cao Bằng saw one Toucan, flown by the commanding officer of GT 1/64 *Béarn*, being shot down.

In October 1947, the first large-scale French anti-guerrilla operation saw three paratrooper battalions being dropped from AAC.1s over Bắc Kạn and Chợ Mới in the Tonkin area. A second Toucan squadron, GT 3/64 *Tonkin*, was formed before the commencement of the operation. This was codenamed Operation Léa, which was an attempt to capture the Vietnamese provisional government, including Hồ Chí Minh (this proved unsuccessful). Other parachute drops were made over Cao Bằng, part of an important supply route from China. GT 3/64 was eventually disbanded in July 1948.

A parade under arms at Bạch Mai in 1948. (*Albert Grandolini*)

On 7 November 1948, Operation Ondine was launched. Escorted by sixteen Spitfires, twenty Toucans and six C-47s dropped paratroopers over Hòa Bình, spearheading the French offensive to regain control over the Red and Black Rivers.

Initially, aircraft losses were light due to Việt Minh's lack of heavy anti-aircraft artillery. Nevertheless, sixteen AAC.1 attrition replacement aircraft were transported to Indochina aboard the aircraft carrier *Dixmude*, leaving France on 25 June 1949. Another Toucan squadron, GT 2/62 *Franche-Comté*, was transferred from Algeria in August 1949 with twenty-four aircraft. GT 2/62 was commanded by a Vietnamese national, Lt Col. Nguyễn Văn Hinh, who was later to become the commander of the South Vietnamese armed forces. Operations included supporting isolated garrisons along the Colonial Route 9 along the Vietnamese-Chinese border. The small airstrip at Cao Bằng was the hub of these operations. During the seasonal monsoon, the airstrip became a sea of mud, with the Toucans being forced to drop part of their load prior to landing to reduce the weight of the aircraft—at least two Toucans were destroyed in landing accidents. Cao Bằng was overrun by Việt Minh guerrillas, despite both Bell P-63 Kingcobras strafing the surrounding areas along with Toucan bombers. In 1951, another airborne assault to reconquer Hòa Bình was launched. Two paratrooper battalions were dropped by twenty-eight Toucans and twelve C-47s, with the landing zone having previously been attacked by B-26 Invaders. Troop reinforcements were brought in later the same day.

AAC.1s were regularly used for casualty evacuation. French nurses, known as Convoyeuses de l'Air, usually accompanied such flights. Many nurses made parachute jumps into combat zones, with several being killed. Night bombing sorties against Việt Minh targets began on 15 October 1951, with AAC.1s of GT.2/62 and NC 701 Martinets of ELA.52 performing such flights. During the siege of Hòa Bình, Toucans were used to deliver supplies to the beleaguered garrison, as well as

The loading of 50-kg anti-personnel bombs onto a Toucan. (*Albert Grandolini*)

The aftermath of a Việt Minh artillery strike against Hòa Bình in January 1952. (*Albert Grandolini*)

evacuating wounded soldiers. In January 1952, Việt Minh artillery accounted for one AAC.1 of GT 2/62 at Hòa Bình, as well as damaging another one. The Battle of Nà Sản in late 1952 saw the final large-scale use of Toucans in Indochina. The 12,000-strong French garrison could only be supplied by air—the idea was to lure Việt Minh forces into battle. In the event, Việt Minh made several failed attempts of conquering Nà Sản, eventually pulling back in December 1952. The victory of Nà Sản gave the French a false sense of security.

The Toucans were also utilised for a very different kind of task. With Việt Minh guerrillas using the regular telephone net in areas under their control, the idea arose of cutting the telephone wires. This was to be accomplished by lowering a hook attached to a rope out of a Toucan flying at extremely low level and literally ripping off the telephone wires. A number of sorties were flown, with great results, though the hook-on-a-rope was later replaced by chain links. On some occasions, the telephone poles were torn from the ground by using this method, along with the telephone wires.

By the early 1950s, the Toucan had been largely withdrawn from Indochina, being replaced by more capable Douglas C-47s. In early 1953, sixteen Toucans remained in service. A US National Intelligence Estimate 91, dated 4 June 1953, stated that fifteen Toucans were in Indochina, including three in Tonkin, eight in Annam, and four in Cochinchina.[1]

Following the French defeat at Dien Bien Phu, negotiations regarding the French withdrawal from Indochina were initiated in Geneva. The ceasefire began on 27 July 1954, with French units finally leaving Vietnam, Cambodia, and Laos the following year.

Three Toucans seen on a primitive landing strip in Vietnam. (*Albert Grandolini*)

Algeria

On 1 November 1954, civil unrest began in Algeria, with the national movement FLN initiating an armed struggle against France. As had occurred in Indochina, AAC.1s were employed as bombers. This was quickly discarded with the arrival of more capable aeroplanes. However, Toucans were heavily used as transports, connecting French garrisons across Algeria.

In July 1955, a new S/GMMTA, which included three air transport squadrons, was formed. Only one of these squadrons, GT 1/92 *Algérie*, used Toucans, with the others being equipped with Douglas C-47s and Nord 2501s. By the end of 1955, GT 1/92 had relinquished its Toucans in exchange for Nord 2501s. Another important task was anti-smuggling patrols, with the majority of such patrols over the Sahara desert flown using AAC.1s of GSRA (*Groupe Saharien de Reconnaissance et d'Appui*) 76 *Tindouf* and GSRA 78, which were part of GOM 86 (literally Overseas Air Group 86). Formed in February 1956, GSRA 76 had an established strength of twenty-four Toucans. Flying across the vast Saharan desert, the job was to detect camel caravans bringing arms and supplies for the FLN rebels. The Toucan was considered ideal for such sorties, being able to carry out reconnaissance, transport, and bombing (when necessary), as well as maintaining regular patrols over the vast desert areas. With the French preparing to set off an experimental nuclear charge at Colomb-Béchar, GSRA 78 was formed in August 1956 for the purpose of surveying the surrounding area, with an air strength of twenty Toucans. Using the Toucan as a bomber ended in late 1957 after one aircraft had suffered heavy damage after being struck by shrapnel from its own bombs.

GSRA 76 had a number of detachments across the Sahara desert, including at Colomb-Béchar, Oran–Tindouf, and Mecheria–Timimoun. Small French garrisons received regular visits from the AAC.1 Toucans of GSRA 76. In the event, GSRA 76 was the final *Armée de l'Air* unit to operate Toucans. Retirement took place in early 1961, with most Toucans being scrapped. Several aircraft were sold to Portugal.

A Toucan seen somewhere in Algeria following a heavy rain shower. (*Albert Grandolini*)

Aéronautique navale

Fifty-three AAC.1s were delivered to the *Aéronautique navale*. The first ten aircraft were ordered in November 1944, with the order soon being increased to twenty-five. With the French road and railway infrastructure largely destroyed or heavily damaged, the requirement for transport aeroplanes was considerable. The intention was to use the AAC.1s for transport and training duties, replacing the ageing Glenn Martin 167Fs then still in use. The *Aéronautique navale* did not refer to their AAC.1s as Toucans, but rather as Ju 52s, Julies, or simply Junkers.

A minimum of three transport squadrons were required, one based in Paris, one based on the French Mediterranean coast, and one in French North Africa. The first *Aéronautique navale* AAC.1, c/n 24, was delivered in February 1945, and by the end of the war five had been delivered. They were used to repatriate French prisoners of war and other detainees liberated from German camps.

In the event, only two squadrons were formed, Escadrille 31.S based near Paris and Escadrille 32.S based in Tunisia—each had nine aircraft, plus three in reserve. A few of the aircraft were gradually transferred to other escadrilles, including 50.S (the Naval Academy), 10.S (a test and experimental unit), and 52.S (an Algerian-based training school, later redesignated 4.S).

In April 1945, an *Aéronautique navale* air transport service was established (Service des Liaisons Aériennes de la Marine, SLAM). Apart from the pair of Ju 52/3m Escadrilles, French-built Dornier Do 24s were also part of SLAM, which operated in similar fashion to a scheduled airline, conducting regular passenger and cargo flights between *Aéronautique navale* bases, both in Metropolitan France and North Africa.

Beginning in 1946, additional Ju 52s arrived. Deliveries included four civilian-registered aircraft, which were taken on charge in 1951. The final aircraft were taken on charge as late as 1958, when several were transferred from Armée de l'Air stocks. Some of the aircraft were former Luftwaffe aircraft that had been refurbished in France post-war.

During the 1950s, Ju 52s of escadrille 11.S were regular visitors to Britain, ferrying *Aéronautique navale* crews that were to collect Avro Lancasters and Short Sunderlands. At this time, many of the Julies were serving in the communications role with overseas units, such as 7.S based in

A rare picture of an Aéronavale AAC.1 in Algeria. (*Albert Grandolini*)

Dakar in West Africa, as well as Madagascar and Indochina. Two or three Julies were also used by 10.S for testing various equipment. Another unit operating Ju 52s was 54.S, which was a fighter training unit.

In 1953, three Julies were attached to a combat unit in Indochina, 28.F, which operated Convair P4Y-2 Privateer bombers. Although *Armée de l'Air* AAC.1s had been operated as *ad hoc* bombers, the *Aéronautique navale* aircraft were only used as light transports. By the late 1950s, the Ju 52s had been relegated to the role of 'station hack'. Larger and more capable transport aircraft, such as Douglas DC-4 and DC-6s as well as Avro Yorks, had taken the place of the venerable Ju 52/3m.

Apart from being operated as a transport, the *Aéronautique navale* also used its Julies for gunnery training and target towing. The former included three aircraft fitted with SAAM AB-151 dorsal gun turrets used as gunnery training platforms in North Africa—plans to fit the PM481 gun turret were not realised. Interestingly, the option to use captured German MG weapons was discussed, but apparently not carried through by the *Aéronautique navale*.

Another training variant was the Ju 52 EBS (*Ecole des bouées sonores*). Two aircraft were modified for use by 12.S in the training of airborne sonobuoy operators for anti-submarine warfare (ASW). These Julies could easily be operated for search and rescue (SAR) duties, while other Julies were used for radio and radar training. Escadrille 56.S, based at Agadir in Morocco, had Ju 52/3ms fitted with full-span spraying gear, being used to combat locusts. The final *Aéronautique navale* unit to operate Julies was escadrille 5.S, based at Karouba, Tunisia. The last Ju 52s were withdrawn from use in 1962.

A radar operator in an Aéronavale AAC.1. Detecting bands of arms smugglers was of vital importance to the French war in Algeria. (*Albert Grandolini*)

Great Britain

Although a sizeable quantity of Ju 52/3ms were assigned RAF serial numbers, few were actually taken on charge. The first two Ju 52/3ms were formally taken on charge on 31 January 1943, being assigned the serial numbers HK919 and HK920. The exact provenance of these aircraft is unknown, but it may be presumed that they had been captured in the Western Desert. HK919 (W Nr unknown) initially served with 173 squadron, based at Heliopolis. In May 1943, the Ju 52/3m was transferred to the British Airways Repair Unit (BARU). Here, HK919 was flown for some seventy hours and forty minutes in twenty-one flights during August alone. By the end of the war, HK919 was on charge with the Middle East Communications Squadron. It was finally scrapped at Heliopolis in August 1945. The second Ju 52/3m, HK920 (W Nr 5005) served with the BARU until being struck off charge on 28 January 1944 after being damaged in an accident. Interestingly, during the autumn of 1943 HK920 had its BMW 132 engines replaced with P & W Wasps—more than likely to ease maintenance.

Following the end of the war, singular Ju 52/3ms were briefly used by different RAF units. Among these were one Ju 52/3m operated by the RAF Austria Communications Flight. Initially, it was planned to transfer the Ju 52/3m to Yugoslavia; however, this fell through due to political considerations. Greece had previously made a request for a number of Ju 52/3ms to be supplied, though this was turned down, thus making it 'politically impossible' to transfer any aircraft to Yugoslavia. Incidentally, the Ju 52/3m made only one confirmed flight with the Austria Communications Flight, a brief air test on 28 February 1946.

One solitary Ju 52/3m was also briefly on charge with 83 Group Communications Squadron (GCS), which operated a heterogeneous mix of liaison and light transport aircraft. Indeed, its first aircraft was a Ju 52/3m, flown by Section Leader Babbage DFM from Schleswig on 15 May. However, the Ju 52/3m had been transferred to what would later become EASSU (see below) by

Two RAF-marked Ju 52/3ms, VM927 and VN743, seen at the Fokker works on 16 April 1946 during refurbishment. (*Thijs Postma*)

the following month. Incidentally, the 83 GCS used many former Luftwaffe Bf 108s, Fi 156s, and Si 204s, but all of these had been withdrawn from use in April 1946.

A few Ju 52/3ms received AM (Air Ministry) serials, including AM 102 (W Nr 6840) formerly D-AKUA of DLH, AM 103 (W Nr 6567) formerly D-AGAC of DLH, and AM 104 (W Nr 641038) formerly D-AUAV of DLH, all surrendered at Flensburg. AM 102 and AM 103 were held at Farnborough before being sold as scrap to J. Dale & Co. Ltd in 1948. Incidentally, AM 104 enjoyed an interesting post-war history. In 1946, the Ju 52/3m was transferred to Miles Aircraft at Woodley. Subsequently, it was donated to the Reading Sky Observers, and was used as a club house. The Reading Sky Observers were soon afterwards renamed Air-Britain, which exists to this day. Air-Britain remains one of the world's best-known aviation historic societies.

Another Ju 52/3m (W Nr 5375) was briefly used by the RAE General Duties Flight, with the serial AM GD1 being assigned. Its last recorded flight took place on 14 July 1945, a return flight from Schleswig to Gilze-Rijen.

Following the end of the war, an additional 103 Ju 52/3ms were taken on charge by the RAF. With the return of American-built transports supplied under Lend-Lease, these Ju 52/3ms would be offered to British airlines for use until indigenous British transport aeroplanes could be delivered. A RAF unit, the Enemy Aircraft Servicing and Storage Unit (EASSU) was formed on 27 September 1945 at Hamburg Airport, Fuhlsbüttel, to handle the former Luftwaffe aircraft. Several Ju 52/3m MS '*Mausi*' were also surrendered, although these were for the most part used for spares recovery. This was due to the fact of the additional work necessary to convert the aeroplane to standard passenger or cargo configuration.

In the event, the RAF did not need the Ju 52/3ms, and very few of them were issued to RAF units. Several of them did not even reach British shores, instead being scrapped at Hamburg Aiport, Fuhlsbüttel, while others were disposed of to foreign governments. Of those that did arrive in Britain, many were sold to Shorts for refurbishment as civilian airliners. A few were assigned to the RAE Transport Flight at Farnborough. These did not last long, though, being withdrawn from use in October 1946. The remaining aircraft were stored at 6 MU at RAF Brize Norton, until sold on 11 March 1948 to scrap dealer J. Dale & Co.

RAF Ju 52/3ms

RAF Serial	W Nr	P/i	Place of Surrender	Date Arrived at EASSU	SoC	Remarks
VM892			Northern Germany	25.10.1945		
VM900	5232	TGr 20	Kjevik, Norway	08.10.1945		Std 6 MU, RAF Brize Norton from 19.11.1945. Scrapped 11.03.1948
VM901	6064	TGr 20	Kjevik, Norway	08.10.1945		Std 6 MU, RAF Brize Norton from 15.01.1946. Sold to Shorts 21.03.1946
VM902	6724	TGr 20	Eggemoen, Norway	08.10.1945		Std 6 MU, RAF Brize Norton from 20.12.1945. Sold to Shorts 21.03.1946
VM903	640758	TGr 20	Eggemoen, Norway	16.10.1945		Scrapped 11.03.1948

VM904	6294	TGr 20	Kristiansand-Kjevik, Norway	08.10.1945	16.07.1946	Destroyed at Fuhlsbüttel 17.07.1946 by 8401 ADW
VM905	3448	Teilkdo 2/ MinensuchGr 1, 3K+KK	Oslo-Fornebu, Norway	08.10.1945		To 6MU 27.11.1945. Scrapped 11.03.1948
VM906	640066	2./TGr 20, 7U+AK	Oslo-Fornebu, Norway	16.10.1946		To 6MU 27.11.1945. Sold to Shorts 21.03.1946
VM907	'729'			12.10.1945	28.02.1946	To Netherlands gvt
VM908	6750	D-APZX	Oslo, Norway *	18.10.1945		R G-AHBP
VM909	640994	TGr 20	Eggemoen, Norway	16.10.1945	16.07.1946	Destroyed at Fuhlsbüttel 17.07.1946 by 8401 ADW
VM910	3443	Teilkdo 2/ Minensuch Gr 1, 3K+BK	Oslo-Fornebu, Norway	08.10.1945	06.06.1946	To 6MU 31.05.1946. Damaged at RAF Manston 01.06.1946
VM911	130740	TGr 20	Kristiansand-Kjevik, Norway	16.10.1945	16.07.1946	Destroyed at Fuhlsbüttel 07.1946 by 8401 ADW
VM912	3003	TGr 20	Kristiansand-Kjevik, Norway	08.10.1945		To 6MU 14.06.1946, scrapped 11.03.1948
VM913	640604	TGr 20	Eggemoen, Norway	16.10.1945		To 6MU 14.06.1946, scrapped 11.03.1948
VM914	5810	TGr 20	Kristiansand-Kjevik, Norway	28.10.1945		To 6MU 07.11.1945, scrapped 11.03.1948
VM915	501195	*Jagdführer* Norwegian	Kristiansand-Kjevik, Norway	28 Oct 45		Sold to Shorts 21.03.1946
VM916	6069	4./TGr 20, 7U+EM	Oslo-Fornebu, Norway	28.10.1945		Sold to Shorts 26.06.1946
VM917	7339	TGr 20	Bardufoss, Norway	28.10.1945		To 6MU 27.11.1945, scrapped 11.03.1948
VM918	501367			05.11.1945		To Leck 27.04.1946 for *Armée de l'Air*
VM919	3366			05.11.1945		To Leck 09.05.1946 for *Armée de l'Air*
VM920	501167			05.11.1945		To Leck 27.05.1946 for *Armée de l'Air*
VM921	501148			05.11.1945		To 6MU 20.12.1945, scrapped 11.03.1948
VM922	3433			12.11.1945		To 6MU 09.12.1945, scrapped 11.03.1948
VM923	501441			12.11.1945		Sold to Shorts 28.02.1946. R G-AHOC

VM924	3436			12.11.1945		To 6MU 05.02.1946, scrapped 11.03.1948
VM925	3345			12.11.1945		To Leck 10.07.1946 for *Armée de l'Air*
VM926	3298			12.11.1945		Sold to Shorts 21.03.1946
VM927	3428			16.11.1945		To Netherlands gvt 46. R PH-UBB
VM928	3376			16.11.1945		To Leck 03.05.1946 for *Armée de l'Air*
VM929	3452			16.11.1945		To 6MU 20.06.1946, scrapped 11.03.1948
VM930	3341			16.11.1945		To Belgian gvt 46. R OO-SNC, ntu
VM931	501220			16.11.1945		To Czech government
VM932	3412			16.11.1945		To Leck 03.05.1946 for *Armée de l'Air*
VM961	641007	TGr 20	Oslo-Fornebu, Norway	21.11.1945		To 6MU 15.01.1946, scrapped 11.03.1948
VM962	641051		Denmark?	23.11.1945		To 6MU 20.12.1945, scrapped 11.03.1948
VM963	052/0097		Denmark?	23.11.1945	16.07.1946	Destroyed at Fuhlsbüttel 07.1946 by 8401 ADW
VM964	3353			21.11.1945		To 6MU 20.12.1945, scrapped 11.03.1948
VM965	501207	409 Repair & Salvage Unit		21.11.1945	16.07.1946	Destroyed at Fuhlsbüttel in 07.1946 by 8401 ADW
VM966	3422			21.11.1945		To 6MU 05.06.1946, scrapped 11.03.1948
VM967	3453			21.11.1945		To 6MU 20.12.1945, scrapped 11.03.1948
VM968	500350			21.11.1945		To 6MU 09.01.1946. Sold to Shorts 21.03.1946
VM969	7334	TGr 20	Kristiansand-Kjevik, Norway	22.11.1945		To 6MU 20.12.1945, scrapped 11.03.1948
VM970	5096	RK+AV	Oslo-Fornebu, Norway	22.11.1945		To Shorts 10.01.1946
VM971	3401	Teilkdo 2/ MinensuchGr 1, 3K+MK	Oslo-Fornebu, Norway	22.11.1945		To 6MU 15.01.1946. Sold to Shorts 03.04.1946
VM972	6780	TGr 20	Kristiansand-Kjevik, Norway	22.11.1945		To Leck 13.07.1946 for *Armée de l'Air*
VM973	3372			22.11.1945		To Leck 12.07.1945 for *Armée de l'Air*

VM974	3433 (same as rep for VM922)			22.11.1945		To 6MU 23.04.1946, scrapped 11.03.1948
VM975	3420			22.11.1945	16.07.1945	Destroyed at Fuhlsbüttel in 07.1946 by 8401 ADW
VM976	3424			22.11.1945	16.07.1946	Destroyed at Fuhlsbüttel in 07.1946 by 8401 ADW
VM977	3309			22.11.1945		To 6MU 09.05.1946, scrapped 11.03.1948
VM978	500136			22.11.1945		To Leck 10.07.1946 for *Armée de l'Air*
VM979	3317			27.11.1945		Sold to Shorts 28.02.1946. R G-AHOG
VM980	5984	SX-ACF, Lw		27.11.1945		Sold to Shorts 07.03.1946
VM981	5024			27.11.1945		To 6MU 03.05.1946, scrapped 11.03.1948
VM982	6466			27.11.1945		To 151 Repair Unit, 49 MU. To 6MU on 21.05.1946, scrapped 11.03.1948
VM983	3429			06.12.1945		To Leck 04.07.1946 for *Armée de l'Air*
VM984	3460			06.12.1945	16.07.1946	Destroyed at Fuhlsbüttel in 07.1946 by 8401 ADW
VM985	5715			07.12.1945		To Netherlands gvt 04.1946. R PH-UBA
VM986	3413			11.12.1945		To 6MU 05.06.1946, scrapped 11.03.1948
VM987	3380			14.12.1945	16.07.1946	Destroyed at Fuhlsbüttel in 07.1946 by 8401 ADW
VN124 *						Serial reported allocated in 10.1945. Presumably in error, as serial originally allocated to Si 204D
VN176			Travemünde-Priwall, Germany	09.05.1946		To 6MU 15.05.1946, scrapped 11.03.1948 *
VN177 *						Serial initially allocated to Bü 181, and then cancelled. Rep as reassigned to Ju 52/3m
VN709				14.12.1945		To 6MU 04.05.1946, scrapped 11.03.1948
VN710				14.12.1945		To Leck 08.07.1946 for *Armée de l'Air*

VN711				19.12.1945		Coded G. Ferried to RAE Farnborough for use by the RAE Transport Flt. Never ToC and noted in Farnborough scrap area on 15.12.1946
VN712	5414			20.12.1945	16.07.1946	Destroyed at Fuhlsbüttel in 07.1946 by 8401 ADW
VN713	5415			20.12.1945		Sold to Shorts 16.03.1946
VN714	7427			20.12.1945		To 6MU 09.05.1946, scrapped 11.03.1948
VN715	5528			20.12.1945	16.07.1946	Destroyed at Fuhlsbüttel in 07.1946 by 8401 ADW
VN716	5023			20.12.1945		To Czech gvt 04.1946
VN717	3441			30.12.1945		To Leck 08.07.1946 for *Armée de l'Air*
VN718	130742	Coded 20		30.12.1945		Ferried to Farnborough 27.06.1946 for use by RAE Transport Flt. Last flight 05.09.1946, noted i scrap area at Farnboroug 15.12.1946
VN719				30.12.1945		To 6 MU 04.04.1946, scrapped 11.03.1948
VN720				30.12.1945		To 6 MU 23.01.1946, scrapped 11.03.1948
VN721				30.12.1945		To 6 MU 23.01.1946, scrapped 11.03.1948
VN722				30.12.1945		To 6 MU 04.04.1946, scrapped 11.03.1948
VN723				02.01.1946		To 6 MU 17.02.1946. Sol to Shorts 01.03.1946. R as G-AHOE
VN724				02.01.1946		To 6 MU 04.04.1946, scrapped 11.03.1948
VN725				02.01.1946		To 6 MU 12.02.1946. Sol to Shorts 28.02.1946
VN726				02.01.1946		To 6 MU 04.04.1946, scrapped 11.03.1948
VN727				02.01.1946		Ferried to Farnborough 17.06.1946 for the RAE Transport Flt. Last flight 12.10.1946 Coxyde-Volkenrode, presumed abandoned at Volkenrode

VN728				05.01.1946		To 6 MU 31.05.1946, scrapped 11.03.1948
VN729				05.01.1946		To 6 MU 23.01.1946. Sold to Shorts 07.03.1946. R as G-AHOF
VN730				05.01.1946		To 6 MU 17.02.1946, sold to Shorts 01.03.1946
VN731				05.01.1946	16.07.1946	Destroyed at Fuhlsbüttel 07.1946 by 8401 ADW
VN740	131150					To 6 MU 12.02.1946, sold to Shorts 01.03.1946. R G-AHOD
VN741	641213			18.01.1946		Sold to Shorts 21.03.1946. R G-AHOL
VN742	2998			18.01.1946		Sold to Shorts 16.03.1946. R G-AHOK
VN743	052/0078			18.01.1946		To Netherlands gvt. R PH-UBC
VN744	641227			18.01.1946		Sold to Shorts 28.02.1946. R G-AHOI
VN745	131159			21.01.1946		To Czech gvt 04.1946
VN746	641364			21.01.1946		Sold to Shorts 28.02.1946. R G-AHOH
VN747	640401	TGr 20	Kristianstad-Kjevik, Norway	23.01.19146		To Belgian gvt 04.1946. R OO-SNA
VN748	131149				16.07.1946	Destroyed at Fuhlsbüttel by 8401 ADW
VN749	6786			24.01.1946		Sold to Shorts 21.03.1946
VN750	130863			24.01.1946		To Belgian gvt 04.1946. R OO-SND
VN751	6527	TGr 20	Bardufoss, Norway	06.02.1946		To Belgian gvt, R OO-SNB
VN752	501191	TGr 20	Bardufoss, Norway	06.02.1946		To 6 MU 15.05.1946, scrapped 11.03.1948
VN753	501402			12.02.1946		To 6 MU 24.05.1946, scrapped 11.03.1948
VN754	180528			12.02.1946		To 6 MU 04.04.1946, scrapped 11.03.1948
VN755	131481			12.02.1946		To 6 MU 17.05.1946, scrapped 11.03.1948
VN756	500138			12.02.1946		Sold to Shorts 28.02.1946, R G-AHOJ
VN907	641198			12.02.1946		To 6 MU 21.05.1946, scrapped 11.03.1948

Greece

Along with the four of Junkers G 24s, the three Ju 52/3ms of *Ellinikí Etairia Enaerion Synhinoion AE* (EEES) were requisitioned into service with the Royal Hellenic Air Force (*Ellinikí Vasilikí Aeroporía* (EVA)) on 2 November 1940; these were used to transport troops and supplies to the front. When Greek and British forces were evacuated to Crete, several of the Junkers airliners were (for obvious reasons) fired upon by Royal Navy warships, with one being damaged. Due to the prevailing chaos during the evacuation of Crete in May 1941, none of the Ju 52/3ms were flown to Egypt. Interestingly, one of the Ju 52/3ms, the former SX-ACF (W Nr 5984) was taken over by the RAF in May 1945 and issued the serial number VM980. Following the end of the war, the Greek government requested the supply of a number of Ju 52/3ms from captured RAF stocks; however, this was rejected.

Hungary

The Royal Hungarian Air Force (*Magyar Királyi Honvéd Légierö* (RHAF)) was a major operator of the Ju 52/3m. With the country having been part of the Austro-Hungarian Empire, Hungary received its independence following the end of the First World War. However, the 1920 Treaty of Trianon meant that around two-thirds of Hungary was ceded to neighbouring nations, along with several million Hungarian nationals. A secret air arm was created, which trained a cadre of pilots and mechanics. Following a treaty signed in the Yugoslavian town of Bled on 22 August 1938, Hungary declared its intention of creating a modern air arm. Tensions with the newly created Slovakian Republic included the dropping of leaflets over areas of Slovakia dominated by Hungarian nationals. On one such flight, on 25 October 1938, a Ju 52/3m escorted by a pair of Fiat CR.32 fighters was lost (this particular Ju 52/3m has not been identified). A brief war between Hungary and Slovakia in March 1939 ensued.

In 1939, three MALÉRT Ju 52/3ms were pressed into service with the RHAF, with additional aircraft being ordered from Germany. Following the German invasion of the USSR on 22 June 1941, Hungary sent an expeditionary force to fight alongside the Germans. A regular link between the RHAF units operating on the Eastern Front and Hungary was maintained in 1941–42 by MALÉRT Ju 52/3ms. By September 1942, all surviving MALÉRT Ju 52/3ms had been transferred to the RHAF. At least one, the former HA-JUA, was used as an aerial ambulance in 1943. In RHAF service, the Ju 52/3ms were initially assigned the serial numbers U.861 to 867, which was later changed to S.101 to 107 (the prefix 'S' meaning Szállítás, i.e. transport). One RHAF unit to operate Ju 52/3ms was the 102/1 Szállitó század (102/1 Transport Squadron), which reportedly had nine aircraft in service in late 1944.

Some twenty-three Ju 52/3ms produced by PIRT began entering RHAF service in early 1944, with the serial numbers S.201 to 223 assigned to these aircraft. With Soviet forces advancing towards mainland Hungary, German forces occupied the country. Apart from operational losses, many RHAF aircraft were destroyed in USAAF strafing raids.

Most of the RHAF Ju 52/3ms were lost in such attacks. On 12 October 1944, two Ju 52/3ms were destroyed by USAAF Mustangs, with another ten RHAF Ju 52/3ms and two Fw 58 *Weihe* light transports destroyed the following day. On 22 March 1945, another seven Ju 52/3ms and one Bü 131 were destroyed in USAAF strafing attack against Vaszar airfield. By end of the war, few, if any, RHAF Ju 52/3ms remained operational. It is unknown if any Ju 52/3ms were operated by the post-war Communist Hungarian air arm, but this is unlikely.

Iraq

In May 1941, a specially formed Luftwaffe unit, *Fliegerführer Irak*, was sent to Northern Iraq in support of an anti-British uprising. All of the aircraft were painted in Royal Iraqi Air Force insignia, and, of the twenty Ju 52/3ms involved in *Fliegerführer Irak*, several were destroyed in strafing attacks by RAF fighters, as described elsewhere.

Italy

The Italian Royal Air Force, *Regia Aeronautica* (RA) operated four Ju 52/3ms. All of these were former Ala Littoria aircraft pressed into service. The four Ju 52/3ms were assigned the serial numbers M.M.60410 to 60413, being used for miscellaneous transport duties, including VIP transport. Between September 1940 and January 1941, I-BIZI (M.M. 60410) was attached to the *Corpo Aereo Italiano* (CAI) expeditionary force based in Belgium for operations against Britain. The Ju 52/3m was used to maintain a regular (three times weekly) link with Italy. The Ju 52/3ms were eventually transferred to Lufthansa.

Italian Military Ju 52/3ms

RA serial	W Nr	P/i	Remarks
M.M.60410	4064	I-BIZI	R D-AIAO
M.M.60411	6083	I-BERO	R D-ASPE
M.M.60412	6710	I-BIOS	R D-ASPI
M.M.60413	6765	I-BOAN	R D-AIAI

Japan

Although no Ju 52/3ms served with the Japanese army or naval air arms, it was believed that a number of Ju 52/3ms, along with several other types of German aircraft, were in operational service. As a result, the code name Trixie was assigned in error in 1942 to the Ju 52/3m by the Materiel Section of the Directorate of Intelligence. One Lufthansa Ju 52/3m had in fact briefly visited the Japanese isles, arriving on 22 May 1939. In August the same year, another Lufthansa Ju 52/3m landed in Tokyo on a connecting flight from Bangkok.

Norway

In February 1940, one DNL Ju 52/3m (LN-DAI) was leased by the Norwegian Navy, serial F.400. However, only the numeral 400 was painted on the aircraft. Although operated by DNL, the Ju 52/3m had been delivered to Norway with a full set of military equipment. The intention was to use the aeroplane as a torpedo bomber, however, it was found that the float-equipped Ju 52/3m had too many struts to easily accommodate a torpedo. It was captured at Horten by German forces on 9 April 1940.

A poor, but interesting photo of a RNoAF Ju 52/3m. Coded Y-AB, this particular aircraft was subsequently sold to Portugal. (*Mats Averkvist*)

Post-war, a fairly large number of former Luftwaffe Ju 52/3ms were transferred to the Royal Norwegian Air Force (RNoAF). Of the 512 Luftwaffe aeroplanes remaining in Norway following the surrender, most were destroyed by 8801 Aircraft Disarmament Wing. However, many of the forty-eight Ju 52/3ms located in Norway were either ferried to Hamburg Airport, Fuhlsbüttel, for potential RAF use, or handed over to Norwegian authorities. Of the latter, eight were formally taken on charge by the RNoAF, with others being broken up due to poor condition. The Ju 52/3ms were divided between three units: No. 20 Transport Squadron (based at Oslo Airport, Fornebu), No. 21 Transport Flight (based at Skattøra using floatplanes), and No. 22 Transport Flight (based at Bardufoss). Aircraft of the latter unit were coded within the N-10 to N-19 range. Incidentally, the No. 20 Transport Squadron derived its name from the Luftwaffe unit TGr 20, which had been based in Norway.

The Ju 52/3ms were operated on internal transport routes, with one (W Nr 5555) being equipped for VIP duties. This particular Ju 52/3m had previously been used by the German Governor of Norway, Joseph Teboven. The Luftwaffe camouflage was apparently retained, with some aircraft being painted white, indicating courier use. Initially, RAF roundels were augmented by painting the rudder in Norwegian colours. RNoAF roundels were applied in late 1945. By February 1946, most of the Ju 52/3ms had been withdrawn from use. Two were overhauled, re-entering service on 7 September 1946. Both remained in service until the spring of 1950, when they were sold to the Portuguese air arm.

Norwegian Military Ju 52/3ms

W Nr	P/i	Toc	Soc	Remarks
2982	NO+IO, 7U+PK 2./T.Gr 20	01.06.1945	12.02.1946	20 TF, to 22 TF 10.08.1945–11.1945. Offered to DNL, rejected and scrapped
5555		11.09.1945	12.02.1946	20 TF, TT 1,325 hrs, of which 93 hrs in RNoAF service
6048	N9+DA Verb.St. Norwegen	14.07.1945	13.02.1946	20 TF, VIP aeroplane.
7229		22.06.1945		21 TF, Transferred to DNL
501198		04.08.1945		21 TF, reserved for Polarfly A/S, ntu

501196		07.09.1946	50	Intended for 22 TF, ntu Y-AC
501219		07.09.1946	50	22 TF, Y-AB
501358		10.08.1945		22 TF, W/o on landing at Bardufoss, 10.08.1945. Engines removed, rest burnt *in situ*

Peru

Following US pressure, the pair of Ju 52/3ms of the German-controlled airline Sucursal Peru, registered as OA-HHA (W Nr 5060) and OA-HHC (W Nr 5272) were seized by the Peruvian government on 31 March 1941. Both were transferred to the *Cuerpo Aeronáutico del Perú* (CAP). The Ju 52/3ms were eventually replaced by US-supplied transport aircraft.

Portugal

In 1936, plans for the expansion of the Portuguese air arm, the *Arma de Aeronáutical Aeronáutica Militar* (AM), were initiated. Ten Ju 52/3m g3es were ordered, all of which arrived in Portugal on 23 December 1936. The Ju 52/3ms had a multi-role capability, being able to operate in both the night bombing and air transport roles. In the night bombing role, the Ju 52/3m carried internal bomb shackles, as well as two machine guns, with a crew of five—two pilots, one navigator/bombardier, one radio operator/gunner, and one flight engineer/gunner. The serial numbers 101 to 110 were assigned, which clashed with a batch of DH 82A Tiger Moths (the Ju 52/3m serials were eventually amended to 201 to 210 in early June 1938).

Initially, these aircraft were rarely used. However, this changed in June 1937 following the arrival of German instructors (two pilots and several mechanics), who initiated a training schedule. Up to this point, the Ju 52/3ms were the largest and most capable aeroplanes acquired by the AM. In early July and August, the Ju 52/3ms were transferred to the *Grupo Independente de Aviação de Bombardeamento* (Independent Bomber Group, GIAB), based at Alverca do Ribatejo. In order to obtain operational experience, the Portuguese government sought permission from the Spanish Falangists (i.e. Franco) for some of their pilots to participate in Ju 52/3m night bombing sorties (following outbreak of Civil War in Spain, the Portuguese government showed diplomatic support for General Franco and his Nationalist cause. A number of Portuguese volunteers were allowed to fight in Spain supporting Franco). In the event, two AM pilots flew two Ju 52/3m sorties against Spanish Republican targets, on 13 and 27 July 1937.

Basing the Ju 52/3ms at Alverca do Ribatejo was a temporary measure, the intention being to transfer them to *Base Aérea* No. 1 at Ota, which eventually took place in 1940. Here, the Ju 52/3ms equipped two night bomber *Esquadrilhas* as part of the *Grupo de Esquadrilhas de Bombardeamento Noturno* (Night Bombing Group, EBN). One Ju 52/3m, serial 207, was assigned to *Base Aérea* (BA) *de Sintra* in 1940. In September 1940, the Ju 52/3m was returned to EBN, with serial 201 being received instead. On 2 February 1941, this Ju 52/3m was destroyed in a hangar fire, along with several Breda Ba 65 attack aircraft (apparently, a party was subsequently held to celebrate the destruction of the much disliked Ba 65s).

With the Ju 52/3m being thoroughly obsolete as a night bomber, their secondary task of personnel and freight carrier was more valuable. Early in the war, Major Humberto da Cruz recommended the use of Ju 52/3ms to deliver personnel and supplies to the AM detachment on the Azores, a distance of 1,610 km (1,000 miles) from Lisbon. This meant that an aerial life line between the Portuguese mainland and the Azores could be maintained. In 1942, five Ju 52/3ms, serials 204, 205, 206, 208, and 210, were transferred to Rabo de Peixe. The unit was called *Esquadrilha de Bombardeamento do Grupo de Esquadrilhas Expedicionárias* No. 1. One of the Ju 52/3ms (serial 210) accidentally caught fire some time in 1943 while parked at Rabo de Peixe, eventually being completely destroyed as no firefighting gear was available. On 12 January 1944, the unit was disbanded and the four remaining Ju 52/3ms were flown back to mainland Portugal.

By late 1947, the strength of the AM included one bomber squadron, equipped with five Ju 52/3ms, based at BA Ota. In 1950, two Ju 52/3m g7es were obtained from the Norwegian Air Force. Both were former Luftwaffe aircraft that had been taken over in May 1945. As a result, Ju 52/3m strength increased to ten aircraft.

With the formation of an independent air arm, the Força Aérea Portuguesa (FAP) in 1952, a new serial numbering system was introduced, with the remaining original aeroplanes being given the serial 6302 to 6309 and the pair of former Norwegian ones 6300 and 6301 respectively. In that year, BA 1 at Sintra received a pair of Ju 52/3ms for the training of navigators and radio operators. New serials were introduced in 1952—6300 to 6308 (original Ju 52s) and 6309 and 6310 (ex Norwegian aircraft).

By the mid-1950s, severe maintenance problems with the BMW 132A-2 engines resulted in replacement engines, Pratt & Whitney R-1340-AN-1 taken off North American T-6 Texans, were fitted. The work was performed by Oficinas Gerais de Material Aeronáutico (OGMA).

In May 1955, a paratrooper unit was formed, with the Ju 52/3ms being put into use as parachute jump platforms. In 1956, serial 6308 was written-off in an accident.

The air transport fleet was further strengthened in late 1960, when fifteen AAC.1s were bought from French Air Force surplus stocks—serials 6311 to 6324. The French-built aircraft differed considerably in detail with the earlier German-built Ju 52/3ms. In FAP service, the AAC.1s were colloquially known as *Ju Argelin* (Ju Algeria). Interestingly, their BMW 132T engines were retained in Portuguese service.

In 1964, the distribution of Ju 52/3m/AAC.1s among FAP units were: BA 1—serials 6304, 6307, 6315, 6316, and 6319; BA 2—serials 6309 and 6320; BA 3—serials 6300, 6303, 6306, 6310, 6312, 6314, 6317, 6318, 6321, and 6324; and BA 5—serial 6311. On 31 August 1969, serials 6300 and 6310 collided in mid-air during a training sortie. Both aircraft were slightly damaged, but the crews managed to get the aircraft down in one piece.

In 1972, the final Ju 52/3m and AAC.1s were withdrawn from use. The longevity of service was partly due to the supply from Spain of CASA 352 spares. Several have been preserved, both in Portugal and at museums around the world.

Portuguese Military Ju 52/3ms and AAC.1s

W Nr	Delivered	S/n	P/i	Remarks
5653	23.12.1936	101, 201		Dbf 02.02.1941
5654	23.12.1936	102, 202, 6302		Presumed scrapped
5655	23.12.1936	103, 203, 6303		Preserved Museo do Ar, Alverca
5661	23.12.1936	104, 204, 6304		Stored Museo do Ar, Alverca

5662	23.12.1936	105, 205, 6305		Presumed scrapped
5664	23.12.1936	106, 206, 6306		Preserved Bodø, Norway
5667	23.12.1936	107, 207, 6307		Presumed scrapped
5669	23.12.1936	108, 208, 6308		W/o 1956
5670	23.12.1936	109, 209, 6309		Preserved Belgium
5671	23.12.1936	110, 210		Dbf at BA.4 in 1942
501219	1950	200, 6300	Lw, NoAF	Stored Museo do Ar, Alverca
501196	1950	201(2), 6301	Lw, NoAF	Stored Museo do Ar, Alverca
291	02.12.1960	6310	*Armée de l'Air*	Presumed scrapped
205	02.12.1960	6311	*Armée de l'Air*	Stored Museo do Ar
366	02.12.1960	6312	*Armée de l'Air*	Presumed scrapped
258	02.12.1960	6313	*Armée de l'Air*	Presumed scrapped
234	10.12.1960	6314	*Armée de l'Air*	Presumed scrapped
005	19.12.1960	6315	*Armée de l'Air*	Preserved Brazil
255	10.12.1960	6316	*Armée de l'Air*	Donated to IWM in 1972. Preserved Poland
048 *	26.11.1960	6317	*Armée de l'Air*	Presumed scrapped
392	19.12.1960	6318	*Armée de l'Air*	Presumed scrapped
1011	1960	6319	*Armée de l'Air*	Presumed scrapped
053	26.11.1960	6320	*Armée de l'Air*	Preserved Germany
325 *	19.12.1960	6321	*Armée de l'Air*	Presumed scrapped
127 *	10.12.1960	6322	*Armée de l'Air*	Presumed scrapped
357	19.12.1960	6323	*Armée de l'Air*	Presumed scrapped
1009	1960	6324	*Armée de l'Air*	Presumed scrapped

* Unconfirmed

Romania

The Royal Romanian Aeronautics (*Aeronautica Regalâ Românâ* (ARR)) was a major operator of the Ju 52/3m, with the initial one entering service in 1938. This aircraft had originally been delivered to Romania in 1932 as a VIP aircraft for the President of the FAI.

On 16 October 1940, LARES and its aircraft were transferred to the ARR. The *Grupul aerotransport militar* (Military Air Transport Group) was then based at Bucharest City Airport, Băneasa airfield. In June 1941, one LARES Ju 52 was transferred to the ARR. Additional Ju 52/3ms arrived from Germany in November 1941, with the serial numbers 1 to 31 being assigned. Two *Escadriles* (Squadrons), Nos 105 and 107 had six Ju 52/3ms each (formal name *Escadrila* xxx *transport greu*, Heavy Transport Squadron). Ju 52/3ms with odd serial numbers were allocated to *Escadrila* 105, odd numbers to *Escadrila* 107. *Escadrila* 106 consisted of the former LARES fleet, also being known as *Escadrila* LARES. The Ju 52/3ms, as well as other ARR transport aircraft, were used extensively in support of the Romanian army's operations on the Eastern Front, including at Stalingrad.

In October 1942, Ju 52/3ms of *Escadrila* 105 made several supply flights to the Caucasus, including Nakhiki, in support of the 2nd Romanian Army Division. The advance against Stalingrad saw the Ju 52/3ms of *Escadrila* 105 being heavily utilized (printed sources mention only five Ju 52/3ms, possibly indicating that one had been lost prior to this date). On 22 November, five Ju 52/3ms delivered food and ammunition to Golovski airfield near Stalingrad. Thousands of Romanian soldiers of five Divisions had been surrounded, and sixty wounded officers were brought out by the Ju 52/3ms. Golovski was captured by Soviet forces during the night of 22–23 November. The following day, the Ju 52/3ms dropped canisters of fuel to the Romanian 1st Armoured Division, which risked being encircled by Soviet troops. A sortie to Tatsinskaya the same day had to be abandoned due to heavy anti-aircraft fire. Three Ju 52/3ms landed at Pitomnik on 4 December. Three days later, on 7 December, three Ju 52/3ms arrived at Pitomnik with much needed supplies. On the return flight, VVS fighters shot down one of the Ju 52/3ms, flown by Aurel Ifrim. The last flight from the encircled Stalingrad area was made on 15 January 1943. In all, three Ju 52/3ms were lost during the siege of Stalingrad.

During the spring of 1943, two Ju 52/3ms collided in mid-air over Melitopol airfield. Although the crew of one of the Ju 52/3ms was able to land safely, the second Ju 52/3m crashed fatally.

On 1 April 1943, fourteen Ju 52/3ms were in service. Ten months later, on 1 February 1944, twenty-one Ju 52/3ms were in service. On 14 April 1944, a major supply and evacuation operation, code named Operation 60,000 was initiated. The intention was to evacuate 60,000 Romanian troops from the Crimean Peninsula. The flights were stopped on 28 April by the German High Command, citing 'higher interests'. However, the evacuation by air recommenced on 11 May. The operation ended three days later when the last airfield was captured by Soviet forces, with a total of 42,190 Romanians evacuated by air. One Ju 52/3m, flown by Radu Gligorz (a former LARES pilot), was lost over the Black Sea—believed shot down by Soviet fighters.

On 10 June 1944, several transport aircraft, including four Ju 52/3ms, were destroyed on the ground at Boteni airfield by marauding USAAF P-38 Lightnings.

After Romania left the Axis in late August 1944, the ARR was reorganised. Many captured Luftwaffe aircraft were incorporated into the ARR, including eleven Ju 52/3ms. At the time, the Escadrilla mixtã aerotransport based at Turnisor, had seventeen aircraft in service, including Ju 52/3ms, Ju W 34s and He 111Es. Fighting against German forces began on 7 September, with the air transport units supporting combat operations. On 20 February 1945, two Rumanian Ju 52/3ms were attacked by a Luftwaffe Bf 109 over Filakovo (Slovakia). One of the Ju 52/3ms, serial 25, was shot down.

On 1 February 1945, twenty Ju 52/3ms were in service. Following the end of the war, the *Escadrila mixtã aerotransport* was based at Piešťany. Subsequently renamed *Flotila 7 aerotransport legãturã*, based at Romeo Popescu airfield near Bucharest. The final Ju 52/3ms continued to serve in dwindling numbers until the early 1960s.

Slovakia

Already in 1942, the Slovak Air Force (*Slovenské Vzdušné Zbrane* (SVZ)) attempted to obtain a number of Ju 52/3ms for use as paratrooper trainers and air ambulances. However, it was not until May 1944 that one Ju 52/3m (W Nr 131484) was delivered. The serial S-1 was assigned to the

Ju 52/3m. It was based at Letisko Tri Duby (Sliač Airport), and used for paratrooper training. At the end of July, a second Ju 52/3m (W Nr 131485), fitted out as a VIP transport, was gifted to the Slovakian leader Josef Tiso.

According to one source, the pair of Ju 52/3ms was allocated to the Flying Transport Group (LDS), based at Vajnory near the capital Bratislava. In July 1944, it was planned to expand the transport element of the SVZ to three flights (Nos 51, 52, and 53) all to be equipped with Ju 52/3ms. This did not happen as German forces began to occupy Slovakia on 29 August 1944. Six days later, a Slovakian insurgency, centred at Letisko Tri Duby (Three Oaks), broke out. The insurgents had about sixty aircraft at its disposal, including the two Ju 52/3ms, although both were subsequently destroyed in a German air raid. On 2 September, an RHAF Ju 52/3m that had flown over Fončarda, near Tri Duby, was intercepted by Master Sergeant František Cyprich, flying an Avia B.534 of the insurgent forces. The Ju 52/3m was subsequently shot down. In the end, the insurrection failed, with many of the surviving aircraft used by the insurgents being destroyed in Luftwaffe raids. On 25 October, the last four airworthy Slovakian aircraft, including one Junkers W 34, was flown east.

South Africa

Following outbreak of war, the eleven Ju 52/3ms of South African Airways (SAA) were requisitioned by the South African Air Force (SAAF). At first, the Ju 52/3ms were assigned to 50 Squadron, with the ZS registration prefix being painted out, with the 'last three' of the registration letters being used as identification serials. The Ju 52/3ms were based at AFS Germiston, being operated by the Bomber Transport Brigade, which later became No. 5 Wing. Regular transport services to East Africa were maintained, ferrying troops and supplies. Although supplanted by Lodestars and Dakotas, the Ju 52/3ms remained in service until the end of the war. During the battle for North Africa in 1942–43, several Luftwaffe Ju 52/3ms were captured in repairable shape. Although SAAF serial numbers were allocated to these Ju 52/3ms, most of them were only used as a source for spare parts.

Following the end of the war, most of the SAAF Ju 52/3ms were withdrawn from use, being worn out and lacking spares support. One of the original South African Airways aircraft, ZS-AJF/661, was briefly tested spraying DDT in Zululand, where the tsetse fly was causing many problems. During the trials, the Ju 52/3m was found to be too big and cumbersome, with the job being assigned to 22 Squadron and its Avro Ansons instead.

One Ju 52/3m, serial 29 (one of the captured Luftwaffe aircraft), was set aside for preservation with the National War Museum at Johannesburg. Due to a lack of space, the Ju 52/3m was displayed outside until the early 1950s. Sadly, due to its poor state, the Ju 52/3m was broken up for scrap along with a Ju 88, an Avro Tutor, a Hawker Fury, and a Westland Wapiti.

South African Military Ju 52/3ms

Serial	Variant	P/i	Name	ToC	SoC	Remarks
660	Mge	ZS-AFD			29.03.1941	Destroyed in strafing attack at Jijiga, Ethiopia, 29.03.1941
661	Msai	ZS-AJF			15.10.1946	Sold to Mr P. R. Kok

662, code B	Mesa	ZS-AJG			24.01.1946	
663	Mesa	ZS-AJH			02.08.1944	
664, code D	Mesa	ZS-AJI			26.10.1946	Sold to Metal Sales (Pty) Ltd
665, code E	Msai	ZS-AJJ			26.10.1946	Sold to Metal Sales (Pty) Ltd
666	Msai	ZS-ALO			15.10.1946	Sold to Mr P. R. Kok
667, code G	Msai	ZS-ALP			26.10.1946	Sold to Metal Sales (Pty) Ltd
668	Msai	ZS-ALR				Written-off 7 miles north-east of Mbeya 08.01.1941
669	Msai	ZS-ALS			26.10.1946	Sold to Metal Sales (Pty) Ltd
670	Msai	ZS-ALU			02.08.1944	
26			*Tobruk*	16.03.1943		Written-off Sauerdale, South Rhodesia, 15.10.1943
27			*Gepp Geschwind*	25.04.1943		
29 W			*Sidi Hanesh.* Later renamed *Dronk Susie*	1943		Std Waterkloof Pres South African War Museum, but sold as scrap 14.09.1954 to Fitzgerald & Co., Johannesburg
30			*Kelibia*	1943		
31			*Bizerta*	1943		
32			*Sidi Daoud*	30.08.1943		
33			*Menzel Termini*	1943		
34			*Cap Bon*	1943		
35			*Tripoli Trix*	08.10.1943		
36			*El Haouiena*	25.07.1943?		
37			*Algiers*	14.01.1944?		
38			*La Sebala*			P/i CQ+HH. Sold at 15 AD 26.10.1946

South Vietnam

Although many aircraft were transferred to the fledgling South Vietnamese Air Force (VNAF) by the departing French, including F8F Bearcat fighters, Douglas C-47 Skytrains, and MS 500 Criquet observation aircraft, no AAC.1 Toucans were included. However, one AAC.1 fuselage was noted at Biên Hòa during the late 1960s in use by the South Vietnamese Army (ARVN) Airborne School as a paratrooper trainer.

Spain

The Junkers Ju 52/3m made a name for itself, being used by both the German unit *Legion Condor* and the Nationalist Air Force as a bomber and transport. According to Hitler, the Junkers Ju 52/3m was the deciding factor in the Franco-led insurrection against the Republican government: 'Franco ought to erect a monument to the glory of the Ju 52. It is this aircraft that the Spanish Revolution has to thank for its victory!' The German Nazi dictator's remark concerned the August 1936 airlift of 7,350 troops and equipment from Tétouan in Spanish Morocco to Seville. This airlift did not, in fact, include any Ju 52/3ms from the beginning. Instead, Fokker F.VII/3ms and Dornier Wal flying boats were initially used until joined by nine Ju 52/3ms during the latter stages of the airlift.

One week after the commencement of hostilities on 18 July 1936, twenty Ju 52/3ms were flown from Dessau via Italy to Seville. The Ju 52/3ms were officially delivered to the Compañía Hispano-Marroquí de Transportes. This was the cover name of a company that had been set up explicitly for the purchase of warplanes and other weapons for the Nationalist rebels. The Ju 52/3ms were immediately set to work on the airlift of troops from Spanish Morocco. Simultaneously, preparations for the first bomb raid were made. On 13 August, during a raid against naval targets, the Republican battleship *Jaime I* was damaged by two 205-kg bombs. The Ju 52/3m was flown by *Flugkapitän* Henke and *Leutnant* Graf Hoyos. Nine Ju 52/3m g3e formed the *Escuadra* B, which was divided into three *Escuadrillas. Escuadra* B was commanded by José E. Rodriguez Diaz de Lecca. On 27 August, the Spanish capitol Madrid was raided for the first time, with the Ju 52/3m g3es returning the following day.

These bomb raids were highly embarrassing for the German legation because at the same time Madrid was being served by civilian-registered Ju 52/3ms of Lufthansa. In Spanish service, the Ju 52/3m became known as *Pava* (*Turkey*).

One of the *Escuadrillas* of *Escuadra* B became the nucleus of 4-E-22 *Escuadra*. However, due to its slow speed and increasing opposition from Republican fighters, the Ju 52/3ms soon began to mainly fly night bombing raids. In August 1937, the Nationalist air arm received an infusion of Ju 52s from the *Legion Condor*. In late 1937, the Ju 52/3ms were transferred to two night bomb groups, 1-G-22 and 2-G-22 *grupos de bombardeo nocturno*. The final Nationalist bomb raid involving Ju 52/3ms occurred on 26 March 1939 against Bélmez, only a few days before Republican forces surrendered. Following Nationalist victory, the surviving 23 Ju 52/3ms were assembled at León. It is estimated that during the Civil War the Nationalist Ju 52/3ms had flown about 5,400 operational sorties, dropping 6,397 tons of bombs, and the total number of flying hours was 13,000. Nine aircraft had been lost, including three shot down, two exploded in mid-air, three destroyed on the ground, and one through desertion (serial 22-53).

In all, twenty-five Ju 52/3ms (s/ns 22-5, -47, 48, 54, -55, -56, -59, -64, -67, -69, -71, -76, -77, -79, -80, -83, -87, -89, -91, -92, -93, -94, -95, -99, and -105) remained in service with the Nationalist air arm in March 1939. On 12 May, a military parade was held at Barajas, with 450 aircraft, including thirteen Ju 52/3ms, on display. By the last quarter of 1939, the Ju 52/3ms were distributed among the following units: 11 *Escuadrilla*, Cabo Juby (four), 114 *Grupo de la 17 Escuadra*, Matacán (11), *Grupo de Estado Mayor*, Cuatro Vientos (two), *Escuela de Observadores*, Málaga (two), *Escuela de Transformacion del Grupo Sur*, Jerez (one), and *Escuela de Vuelo sin Visibilidad*, Matacán (four). On 17 November 1939, serial 22-94 was damaged beyond repair in a landing accident at Cabo Juby.

The heterogeneous mix of aircraft inherited from the Civil War was difficult to sustain. Lack of spares ensured that many aircraft remained grounded. In late 1939, plans for a future Spanish Air Force envisaged 5,000 aircraft, including 290 transports. Standardising on a few types of aircraft was a necessity, with the Ju 52/3m being selected as the standard transport aeroplane.

On 29 February 1940, the 114 *Grupo* was disbanded, with the Ju 52/3ms being transferred to the other units listed above. Later on, Ju 52/3ms also served with the *Academia General del Aire*, San Javier, *Escuela des Transmisiones*, Cuatro Vientos, *Escuadrilla del Cuartel General de la Región Aérea Atlántica, Villanubla*, and *Escuadrilla del Cuartel General de la Zona Aérea de Marruceos*, Sania Ramel, later transferred to Tétouan.

Following the German invasion of the USSR on 22 June 1941, a voluntary Spanish air unit was formed. Named *Legion Azul* (Blue Legion), a few Ju 52/3ms were operated for transport and general communication duties. During the war, EdA aircraft rarely ventured away from Spain, with one exception occurring on 13 May 1943, when fifteen Spanish fighter pilots arrived at Vélizy-Villacoublay, France, courtesy of an EdA Ju 52/3m. The pilots were to collect fifteen former Luftwaffe Messerschmitt Bf 109F-4s.

In 1944, the first of an eventual total of 170 CASA 352s began to arrive. In December 1945, the designation T.2 was assigned to the Ju 52/3m and CASA 352. The aircraft already in service received new serials, e.g. 22-5, which became T.2-5, and the last Ju 52/3m in EdA service was T.2-83 (the former 22-83). Originally taken on charge in December 1936, the aircraft was transferred to the *Escuela de Transmisiones a la Maestranza* on 15 June 1959. However, due to wear and tear, the Ju 52/3m was struck off charge on 15 February 1960.

More than sixty EdA units operated the CASA 352. Apart from regular transport squadrons, including the 112, 361, and 362 *Escuadrillas*, various EdA schools and academies also operated CASA 352s. Many fighter squadrons also had at least one CASA 352 for general transport duties. The highest number of CASA 352s in active service was 140, on 31 January 1957. Between 31 January

A five-aircraft formation take-off by EdA Ju 52/3ms. (*Luis Gonzalez Pávon*)

CASA 352 T.2B-252. Note that the upper part of the fuselage has been painted white to reduce heat inside the cabin. (*Luis Gonzalez Pávon*)

1956 and 30 June 1975, no less than 185,923 flight hours were accumulated. It is likely that the total number of flight hours was more than 200,000.

The unit to maintain the CASA 352 in service the longest was the *Escuela Militar de Paracaidistas* (Parachute Training School). Established on 24 July 1947 at Alcantarilla, the unit initially had one solitary Savoia-Marchetti SM 81 for parachutist training. In January 1948, CASA 352 serial T.2-125 arrived, with the first parachute training sortie using the new aircraft occurring on 15 January. The first parachutist to exit the CASA 352 was *Capitán* Joaquín Echevarría Bengoa. The final EdA CASA 352, serial T.2B-181, was withdrawn at Alcantarilla on 4 October 1978. In all, 51,546 paratroopers had been trained during more than 400,000 sorties.

Ifni Campaign

Following the declaration of independence by Morocco in 1956, attempts were made to oust the Spanish from the Ifni enclave in the southern part of the country. Spain responded by sending a contingent of ground and air forces. On 23 October 1957, the Ifni campaign began, ending on 30 June 1958. Although the CASA 352s were regularly employed in the airlift, paratroop drop, casualty evacuation, and liaison roles, experimental bomb dropping trials were also made. This involved CASA 352 serial T.2B-235 (code 36-12) dropping a 200-litre fuel barrel over Tlata de Sbuia in Spanish Morocco. Visually, the result was spectacular, with actual damage being very limited. During the Ifni campaign, EdA CASA 352s flew about 170 operational sorties.

The last remaining CASA 352s were finally withdrawn from use in 1977–78. The Ju 52/3m and CASA 352 had served in Spain for more than forty-two years. Many CASA 352s have been preserved around the world.

A CASA 352 seen during the brief Ifni campaign. Note the pair of camels in the foreground replenishing their water supply. (*Luis Gonzalez Pávon*)

Sweden

Beginning in September 1939, five ABA Ju 52/3ms were pressed into service with *Flygvapnet* (Swedish Air Force). The Ju 52/3m was assigned the designation Tp 5, i.e. Transport Aircraft Type 5. The aircraft were leased for various periods, all being returned to ABA by the end of the war. Apart from regular transport duties, the Ju 52/3ms were also used as a flying lecture hall for the training of navigators. The Tp 5s were attached to different *Flygvapnet* Wings as required. During the spring of 1941, the Swedish High Command planned to occupy the Finnish Åland Islands. A number of impressed civilian transport aircraft, including the ABA Ju 52/3ms, were to carry troops and supplies; however, the operation was cancelled.

In early 1945, the private company AB *Industridiesel* offered two former Luftwaffe Ju 52/3ms to *Flygvapnet*—the asking price was 75,000 kr. Both Ju 52/3ms (codes NO+IJ and 7U+AM) had been interned in Sweden in 1943. However, due to difficulties in establishing formal ownership, the offer had to be declined. The Ju 52/3ms remained in storage and were scrapped in 1946 and 1949 respectively.

Swedish Military Ju 52/3ms

Serial	W Nr	P/i	Leased	Remarks
	4017	SE-ADR	04.09.1939–15.09.1939, 12.04.1940–21.06.1940, 03.06.1941–20.08.1941, 20.09.1941–10.10.1942, 20.10.1942–29.03.1945	Coded TP-2 May–June 1940. 1,310 flight hours accumulated in *Flygvapnet* service
907	5614	SE-AFA	12.04.1940–02.10.1940, 01.11.1940–10.03.1941, 15.05.1941–20.08.1941, 20.09.1941–20.12.1941, 07.01.1942–10.10.1942, 20.10.1942–11.03.1944, 19.05.1944–12.09.1944	Coded TP-1 in 1940

During the war, the ABA Ju 52/3ms were leased by *Flygvapnet* on several occasions. This is SE-AFD. (*Arlanda Civil Aviation Collection*)

909	5620	SE-AFB	12.04.1940–05.10.1940, 26.07.1943–11.09.1943	Coded TP-4 in 1940
908	5633	SE-AFC	12.04.1940–02.08.1940, 04.09.1944–19.09.1944	Coded TP-3 in 1940
	5646	SE-AFD	30.07.1941–27.08.1941, 01.11.1941–11.06.1945	

Switzerland

In early 1939, three Ju 52/3m g4es were ordered by the *Schweizerische Flugwaffe* (Swiss Air Force). Upon delivery, the serial numbers A-701, A-702, and A-703 were assigned. The aircraft were not to be used primarily as transports, but rather as flying lecture halls, training observers for the light bomber squadrons flying the indigenous EKW C-3603/04 combat aeroplane. The cabin of each Ju 52/3m had room for eight trainees and four crew, including pilots and instructors. Gunnery training could also be accomplished, with one dorsal and one lower fuselage gun turret being fitted. The observer trainee stations could be removed, and replaced by standard passenger seats. All three Ju 52/3ms were based at Dübendorf.

Post-war, the Ju 52/3ms were civilian registered to allow flights outside of Switzerland. The purchase of 130 P-51D Mustangs from USAF surplus stocks saw the aircraft flying on a regular basis between Oberpfaffenhofen in Bavaria and Dübendorf. Several shuttle flights took place between 6 September 1948 and 22 December 1948 using A-701/HB-HOS and A-703/HB-HOP. Following the Swiss selection of the DH 100 Vampire for production under licence, the Ju 52/3ms were used to carry DH Goblin engines and other components from Hatfield to Emmen. All three Ju 52/3ms were used, including A-702/HB-HOT. Due to the lack of de-icing gear as well as IFR (both being an essential requirement when flying in typical foggy and inclement English weather), this job was subsequently transferred to Swissair and their more modern DC-3s. In early 1951, severe snowfall caused many deaths in the alpine areas of Switzerland, with many villages becoming isolated. The Ju 52/3ms carried supplies, including food and medicines, as they were the only aircraft available for such tasks.

When the final EKW C-3603/04 two-seat light bombers were withdrawn from use in 1956, the requirement for observers vanished. As a result, the Ju 52/3ms were subsequently relegated to passenger and cargo transport duties, and to train parachutists. The latter task was considerably eased when the entrance door of the Ju 52/3ms was modified to hinge inwards. During the 1950s, the supply of tyres began to run out. In the event, AAC.1 wheel hubs, which had long since been modified to be used with DC-3 tyres, were obtained. On 3 September 1959, the Ju 52/3ms were removed from the civilian register. One interesting event was the cameo role for s/n A-702 as a winter-camouflaged Luftwaffe Ju 52/3m in the Hollywood movie *Where Eagles Dare*. It also appeared in a German tv-series, *Zwei himmlische Töchter* (*Two Heavenly Daughters*).

The Flugwaffe Ju 52/3ms were finally withdrawn from use in 1982. They were the last operational military Ju 52/3ms anywhere in the world, but, instead of being grounded, the Ju 52/3ms flew on, being transferred to the society *Verein der Freunde des Museums der Fliegertruppe*. Later on, the private company Ju Air was formed to operate the trio of Ju 52/3ms (for more details, see Chapter 17 and 19).

Flugwaffe Ju 52/3m g4es

Serial	W Nr	ToC	SoC	Remarks
A-701	6580	10.1939	1982	R as HB-HOS between 13.09.1948 and 03.09.1959. R on 26.08.1982 as HB-HOS
A-702	6595	10.1939	1982	R as HB-HOT between 27.03.1951 and 03.09.1959. R in 07.1985 as HB-HOT
A-703	6610	10.1939	1982	R as HB-HOP between 09.05.1947 and 03.09.1959. R on 15.10.1982 as HB-HOP

Syria

Seven AAC.1s were reportedly acquired in the late 1940s by the *Al Quwwat al-Jawwiya al Arabiya As-Souriya* (Syrian Air Force). However, only two have been positively identified, the former LR-AAI and -AAJ of the Lebanese airline CGT, both being taken on charge on 9 July 1949. The third CGT AAC.1, LR-AAC, may also have ended up in Syria. Presumably, the AAC.1s were operated on regular troop carrying and parachute training tasks until the early 1950s, eventually being replaced by second-hand Douglas C-47s and Russian-supplied Ilyushin Il-14 Crates.

USA

Although only one Ju 52/3m was operated by the USAAF, this particular aeroplane enjoyed a very diverse history, deserving a more detailed description. This particular Ju 52/3m was allocated the designation Junkers C-79, and the serial number 42-52883, being taken on charge on 9 May 1942.

A condensed pre-USAAF history of the Ju 52/3m begins with its manufacture at Dessau. Originally registered as D-AENF, it was delivered to Lufthansa Südamerika on 1 April 1937. After being shipped across the Atlantic, it was used for routes across the Andes, being named *Aconcagua*. The Ju 52/3m was used on many 'show-the-flag' flights across the vast continent. Soon after the war broke out, the Ju 52/3m transferred on 11 September 1939 to Syndicato Condor, and registered as PP-CBA. Just over a year later, on 29 November 1940, it was registered as OA-HHD for Lufthansa Sucursal Peru, subsequently being leased to Sociedad Ecuatoriana de Transportes Aéreos (SEDTA) on 25 April 1941 as HC-SAD. Apparently, the Ju 52/3m and its German crew played an important role prior to the outbreak of war between Ecuador and Peru, making several reconnaissance flights. Following strong US pressure, the aircraft was confiscated on 5 September 1941 by the Ecuadorian government, being transferred to the *Fuerza Aérea Ecuatoriana*.

On 13 May 1942, it was formally acquired by the USAAF. The Ju 52/3m was flown to Albrook Field in the Canal Zone by Lt Ned L. Kragness. Prior to landing, Kragness was met by a pair of USAAF Curtiss P-40 fighters, which provided escort for the final leg. The Ju 52/3m was subjected to a major refurbishment, which included the replacement of the BMW engines for Pratt & Whitney R-1690-23s. It made at least one visit to Aruba, on 9 June 1942, which may or may not have been connected with the hunt for German submarines operating in Caribbean waters. In November 1942, it was assigned to the 20th Troop Carrier Squadron (TCS) of the Sixth Air Force, and based at Howard Air Force base in the Panama Canal Zone.

US aviation historian Dan Hagedorn refers to the solitary USAAF C-79 as being a 'white elephant', noting that the aircraft spent more time in maintenance than in actual service. Apart from the engines, the 'motor mounts, wheels, brakes, instruments, radio and controls' were exchanged during its eighteen-month service with the USAAF. On 6 October 1943, the Ju 52/3m, having undergone yet another sojourn in the maintenance shop, was declared 'awaiting delivery

The sole USAAF Junkers C-79. (*Col Kenneth E. Marts via Dan Hagedorn*)

to station'. However, the Commanding Officer of the 20th TCS stated his obvious disapproval of having the C-79 returned to the unit, with the aircraft later being described as 'unassignable'. In the event, the C-79 was sold to the US Public Roads Administration for $10,000, being delivered on Christmas Eve 1943. The Ju 52/3m was registered in Costa Rica as TI-60 (see page 105 for further details).[2]

It is likely that a number of Ju 52/3ms were found within the US zone of occupation, though none were taken on charge. However, during the summer of 1945, one Ju 52/3m MS surrendered at Schleswig to British forces was transferred to the USAAF. The USAAF had requested the transfer of a number of Luftwaffe aircraft found in the British zone of occupation for intelligence purposes. Details of such aircraft were forwarded to Colonel Harold E. Watson, USAAF. Watson was in charge of Project Lusty, which involved the recovery and transfer to USA of Luftwaffe aircraft, engines, and other equipment of interest for evaluation. Among these were several Ar 234 and Me 262 jets. The American personnel involved were informally known as 'Watson's Whizzers'. The reasoning behind the transfer of the Ju 52/3m presumably belonged to it being equipped for anti-magnetic mines operations. No result of the evaluation of the anti-mine gear is known.

USSR

In 1937, a Ju 52/3m that had come down intact behind Republican lines was shipped to the USSR. Here, it was thoroughly tested at NII VVS. The test pilot P. M. Stefanovsky made no less than 193 flights in the Ju 52/3m.

After the Battle of Stalingrad, many Luftwaffe Ju 52/3ms were found at Pitomnik and other airfields. Although the majority of these had been damaged beyond repair—either through strafing attacks or being deliberately destroyed by the retreating Germans—several were repaired and taken on charge. It is believed that as many as fifty Ju 52/3ms entered service with the VVS. They served into the immediate post-war years, eventually being replaced by more capable transport aeroplanes.

Post-war, one Ju 352A was briefly employed as an engine test bed.

Yugoslavia

In October 1944, one Luftwaffe Ju 52/3m was captured intact at Nikšić, Montenegro. The Ju 52/3m was subsequently incorporated into service with the *Narodnooslobodilačke Vojske Jugoslavije* (Headquarters of the People's Liberation Army of Yugoslavia, NOVJ). Soon afterwards, the crew of a second Ju 52/3m defected to Zadar. Both Ju 52/3ms were transferred to the Yugoslav Air Force (JRV) on 2 April 1945. It is likely that the pair of Ju 52/3ms received the serial numbers 7201 and 7202.

In 1947, six second-hand AAC.1s were acquired by the JRV, with the serials 7203 to 7208 being assigned. Some of these had been delivered to Yugoslavia already in August 1945, and the last of these AAC.1s, serial 7208, was the former F-BBYB. At least two AAC.1s/Ju 52/3ms were lost in JRV service, one on 29 June 1951 at Mrzle Vodice and one on 7 January 1956 at Pelješac. The remaining AAC.1s were withdrawn from use in 1964, and only one, s/n 7208, has been preserved.

19

Preserved Junkers Ju 52/3ms

A fair amount of Ju 52/3ms have been preserved, including post-war AAC.1s and CASA 352s. In all, some fifty aircraft survive. Several wrecks remain in remote, inaccessible locations, particularly in northern Scandinavia, with others found underwater (including one that was recently located in the Black Sea and one in a lake in Germany). With the Ju 52/3m being fairly common among preserved vintage aeroplanes, it is unlikely that more airframes will be added as the cost of recovery and restoration make such ventures unlikely. The following list contains information on complete airframes on display or under restoration, as well as substantial components by country in alphabetical order. Also included are details on the eight airworthy Ju 52/3ms and CASA 352s. Regarding the latter, the Swiss company Ju Air operates three Ju 52/3ms and one CASA 352. The Lufthansa *Traditionsflug* has one Ju 52/3m, obtained from the USA in 1986, with one former Doug Arnold aircraft flying in France. One CASA 352 is still flying in South Africa, as is one CASA 352 owned by the Military Aviation Museum, Virginia Beach, USA.

On 24 February 1943, this Ju 52/3m of 4./KGr z b V 108, coded 7U+LM (W Nr 6032), crashed in Northern Norway due to poor weather. The crew all sustained various injuries, and were extremely lucky to have survived the impact. The Ju 52/3m was abandoned due to the inaccessibility of the crash site—this is how it looked in 1980. Since then, some parts have been recovered by the Bodø Air Museum. (*Pär Erixon*)

Argentina

Museo Nacional de Aeronáutica, Morón

W Nr 4043
D-3356
D-ABIS
PP-CAX
T-158
LV-ZBD
LQ-ZBD

When the final Ju 52/3ms were withdrawn from use in the early 1960s, one was set aside for preservation at the Museo Nacional de Aeronáutica, Buenos Aires. Originally delivered to Lufthansa as D-3356, the Ju 52/3m was soon transferred to Syndicato Condor. Subsequently sold to the Fuerza Aérea Argentina (Argentinian Air Force) in 1946, it was assigned the serial T-158 and named *Yapeyu*. The Ju 52/3m was later transferred to the Ministry of Agriculture, and was converted to an anti-locust sprayer. Registered as LQ-ZBD, it was finally withdrawn on 14 August 1961, having accumulated 14,892 flight hours. It was formally transferred to the Museo Aeronáutico at Aeroparque (later moved to Morón) on 18 July 1963. Remaining on outside display for many years, this early model Ju 52/3m was severely damaged by high winds in 2011, but is under restoration at Rio Cuarto. The Ju 52/3m has been exhibited both as LQ-ZBD and later as T-158, and, during the late 1960s, as 'T-159'. However, the identity of the Ju 52/3m has been established beyond doubt as being W Nr 4043. It is likely to be the oldest complete Ju 52/3m in existence.

Austria

Verein der Freunde historischer Luftfahrzeuge e.V., Mönchengladbach

W Nr 6888

In 2001, the remains of a Ju 52/3m surfaced at the Umblakes glacier, Osttirol, Austria. The investigation that followed showed it to be a Ju 52/3m g6e of KGr z b V 9. Originally delivered to the Luftwaffe on 17 May 1940, the aeroplane had come down on the glacier in the summer of 1941 with the wreck soon being completely covered by ice. The cockpit section and one of the engines was subsequently recovered and restored by Austrian-based Sandy Air Corp. The cockpit is currently on long-term loan to the *Verein der Freunde historischer Luftfahrzeuge e.V.* It is displayed at the Hugo Junkers Hangar at Mönchengladbach airport. Long-term plans call for the Ju 52/3m to be displayed in a purpose-built museum at the glacier where it crashed.

Belgium

Brussels Air Museum, Brussels

W Nr 5670

The cockpit of Ju 52/3m g6e W Nr 6888 seen at Mönchengladbach, Germany, on 21 June 2015. It is intended to be displayed at a yet-to-be-opened museum close to the crash site in the Austrian Alps. (*Author's Collection*)

AM 109
AM 209
FAP 6309
'OO-AGU'

One Ju 52/3m was obtained from Portugal by the Brussels Air Museum in the early 1980s. Originally delivered to Portugal before the war for use as a night bomber, it was assigned the serial number 109. Its military serial was subsequently changed to 209 and then to 6309. After being withdrawn from use, it was one of a large number transferred to the *Museu do Ar*. The Belgian national airline SABENA had operated a number of Ju 52/3ms in the late 1930s, with the surviving aircraft continuing in use in Belgian Congo following the German occupation of Belgium. In 1945, a few Ju 52/3ms transferred from RAF stocks were used for a variety of tasks by the Ministry of Communications.

Volunteers from the airline SABENA expressed an interest in restoring the Junkers to airworthy condition. However, this did not happen, partially due to lack of funds. Instead, it was stored at Zaventem before being moved to Brustem (Sint-Truiden Air Base), where a restoration to static condition was carried out. The tailplane reportedly comes from FAP serial 6310. The Ju 52/3m was finished in pre-war SABENA colours and marked as 'OO-AGU', remaining on display.

The former Portuguese Ju 52/3m preserved at the Brussels Air Museum. (*Miro Patocka*)

Brazil

Museu Asas de um Sonho (Wings of a Dream Museum), São Carlos, Brazil

C/n 005
F-BAJE
Armée de l'Air
FAP 6315

Established in 1996 by Rolim Adolfo Amaro, founder and president of the airline TAM, the *Museu Asas de um Sonho* is adjacent to the TAM Technology Centre at São Carlos airport. In 2006, the museum was opened to the public. One early production AAC.1 was obtained from Portugal in an exchange deal in 2011 with an A-26 Invader going the other way. The AAC.1 is currently under restoration. The TAM Museum has recently closed.

Canada

Western Canada Aviation Museum, Winnipeg

C/n 39
T.2-148
T.2B-148
N99234
CG-ARM

As briefly mentioned, only a few components of the original Ju 52 CF-ARM survived into preservation. The importance of CF-ARM to the development of Canadian civilian aviation meant that the Winnipeg-based Western Canada Aviation Museum initiated a search during the late 1970s for a Ju 52/3m that could be converted into a single-engined Ju 52. One CASA 352 was subsequently located at Orlando, Florida. Owned by the Wings and Wheels Museum, the CASA 352 was duly purchased in December 1981 for $75,000, with James Richardson Ltd, which had funded the purchase of the original CF-ARM, graciously providing a grant.

This particular CASA 352 had originally been delivered to the EdA as T.2B-148. After being withdrawn from use, it was sold to Sir William Roberts in July 1975, who operated a private aviation museum at Strathallan, Scotland. The CASA 352 was soon sold to former USAF fighter pilot Dolphin Overton. Registered as N99234 in 1977, it was flown across the Atlantic two years later in seven legs. Following arrival in Pennsylvania, the centre Beta engine—which had caused problems during the entire flight—was replaced with a Pratt & Whitney R-1340 prior to the final flight to Orlando. When the Wings and Wheels Museum went bankrupt in 1981, the entire collection was put up for auction. Although the CASA 352 was not sold during the auction, a deal was later struck with the Western Canada Aviation Museum.

The old transport was prepared for the long ferry flight northwards, with considerable assistance being provided by the Confederate Air Force and the Canadian Armed Forces. The venture had been named Operation Tante Ju. With the original registration being unobtainable, the CASA 352 was registered as CG-ARM. With several stops along the way, CG-ARM attracted considerable interest, with one American onlooker referring to it as a 'funny looking cigarette box!' Following arrival at Winnipeg, CG-ARM was flown to Gimli for storage. In 1983, the Western Canada Aviation Museum obtained permanent residence at Winnipeg International Airport. The CASA 352 was removed from storage and flown to Winnipeg, also the last time it would ever fly. Restoration to single-engined status began in earnest, with the work being led by 'Sandy' A. Hyland. The outer Beta engines were removed, and a Rolls-Royce Merlin fitted *in lieu* of a Buzzard. A four-bladed wooden propeller was also fitted. The cabin windows were reconfigured, as was the cockpit. Restoration and conversion work by museum volunteers lasted 3,500 hours. In 1985, the 'new' CF-ARM was placed on display at the Western Canada Aviation Museum.

CASA 352 CG-ARM seen during its ferry flight from Orlando to Winnipeg. (*Robert S. Grant*)

Above: Restoring the three-engined CASA 352 into a replica of the single-engined Ju 52 ce CF-ARM took nearly three years. (*Larry Milberry*)

Opposite above: In 1988, this Ju 52/3m fuselage was traded by the *Vojenské Muzeum* for a Tiger Moth. Today, it can be seen at Laatzen. (*Lukas Syvory via Miro Patocka*)

Opposite below: A Ju 52/3m fuselage section is held by the *Vojenské Muzeum*, Kbely, Czech Republic. (*Lukas Syvory via Miro Patocka*)

Colombia

W Nr ?
FAC 625

One early production Ju 52/3m, FAC serial 625, is on outdoor display at the FAC Museum, El Dorado, Bogotá.

Czech Republic

Vojenské Muzeum, Prague-Kbely

W Nr ?
OK-xxx

A 3-metre section of the centre fuselage of a Ju 52/3m is held by the *Vojenské Muzeum*. Faint traces of the CSA logotype can be seen on the fuselage, which remains in storage. Interestingly, a relatively complete Ju 52/3m fuselage was exchanged in 1988 by the *Vojenské Muzeum* for DH 82A Tiger Moth D-EBIG (the former PG627). The fuselage had been discovered during the 1970s in the town of Nymburk, where it was used as a shelter by forestry workers. The fuselage is currently on display at the *Luftfahrtmuseum Hannover-Laatzen*.

Finland

Although no complete Ju 52/3m can be found in Finland, components of two Luftwaffe aircraft are preserved.

Keski-Suomen ilmailumuseo, Tikkakoski, Finland

W Nr 5596
Lw 7U+OM

W Nr 6810?
Lw 7U+OK

In 1973, the wreck of a former TGr 20 Ju 52/3m, 7U+OM (W Nr 5596) was reportedly to be recovered by the Finnish Air Force Museum. It had been shot down on 19 October 1944 by a Soviet Pe-3. In the event, the whole aircraft was not recovered, and therefore only one engine with propeller as well as a section of the fuselage, showing the code OM, was placed on display at Tikkakoski. The left engine is exhibited at Inari. Also held by the museum is the tail section of another TGr 20 Ju 52/3m, 7U+OK (possibly W Nr 6810). Shot down over Lake Muddusjärvi on 21 October 1944, the wreck was discovered in 1959. The Ju 52/3m was raised from the lake, but scrapped four years later. However, the tail section remained in the lake until 2003 when it was recovered. The tail section is privately owned, being kept at Ivalo outside a Sotka store.

France

One indigenous AAC.1 and one CASA 352 survives in France, with the latter being airworthy.

Musée de l'Air, Le Bourget, Paris

C/n 216
C/n '334'

Only one AAC.1 survives in France, this being c/n 216, preserved at the *Musée de l'Air*. It was marked as c/n 334 and coded DG in honour of the Director of the *Musée de l'Air*, who had flown that particular AAC.1. A former *Aeronavale* aircraft, it is currently on loan to the *Musée de Tradition l'Aeronautique Navale* at Rochefort-Soubise.

The *Musée de l'Air* AAC.1 seen during the late 1970s. (*Swedish Aviation Historical Society*)

Amicale Jean-Baptiste Salis, La Ferté-Alais

C/n 24 or 103
T.2B-212
G-BECL
F-AZJU

In 1990, CASA 352 G-BECL was acquired from Aces High by *Amicale Jean-Baptiste Salis*. The trimotor arrived at La Ferté-Alais on 1 June, with a team of twenty volunteers being assigned the task of restoring the aircraft. During this work, two data plates, dated 1943 and 1946, indicated that the airframe had been built in Germany as part of a batch of thirty (c/n 1-30) delivered to Spain during the final stages of the war. Many German-built components, including the fuselage, were evident, resulting in the aircraft being officially registered with the French civil aviation authorities as a Ju 52/3m. Twelve years and 19,000 hours of restoration work later, the aircraft took to the air once again. Registered as F-AZJU, the Ju 52/3m is currently based at La Ferté-Alais, visiting air shows around France and continental Europe during the summer months.

Germany

Unsurprisingly, a fairly large number of Ju 52/3ms—including AAC.1s and CASA 352s—are found in Germany, including one airworthy aeroplane. The fuselage of an unidentified Ju 52/3m is on display at the *Luftfahrtmuseum Hannover-Laatzen*.

Lufthansa *Traditionsflug*, Hamburg

W Nr 5489
D-AQUI
LN-DAH
D-AQUI
LN-KAF
HC-ABS
N130LW
N52JU
D-CDLH (marked as 'D-AQUI')

The best-known Ju 52/3m in Germany is 'D-AQUI', which is airworthy with Lufthansa *Traditionsflug*. Originally delivered to Lufthansa on 10 April 1936, it was fitted with floats and operated on the Copenhagen–Gothenburg–Oslo route. However, the Ju 52/3m was sold to DNL on 1 July 1936, and registered as LN-DAH. Following the German occupation of Norway, the aeroplane found its way back to Lufthansa, courtesy of the Luftwaffe. Re-registered as D-AQUI in 1940, the Ju 52/3m was used on the Norwegian routes, connecting Trondheim in the central part of the country to Kirkenes on the Russian border. It was one of many Ju 52/3ms left in Norway following the German surrender.

After being overhauled, the Ju 52/3m was briefly operated by the Norwegian Air Force before being transferred to DNL. Severe corrosion saw the fuselage being replaced with that of a former Luftwaffe Ju 52/3m, W Nr 130714, and the starboard wing was also replaced. Registered as LN-KAF, the Ju 52/3m was transferred to the new Scandinavian airline SAS, remaining in use until 1956 when it was sold to Ecuadorean airline TAO. On 30 July 1957, the Ju 52/3m was loaded aboard M/S *Margrethe Bakke*, along with some 20 tons of spare parts. Following arrival in Quito in August 1957, the Ju 52/3m was registered as HC-ABS. During the following five years, HC-ABS saw extensive use before lack of spares and the need for a major overhaul finally grounded the old Ju 52/3m. Instead of being scrapped, the Ju 52/3m was towed to a quiet corner of the airport, and abandoned to the elements.

Rescue and resurrection arrived in the shape of Lester Weaver of Polo, Illinois, USA. Weaver purchased the derelict Ju 52/3m on 25 May 1970 for $5,000. It was rebuilt, registered as N130LW, and ferried to the USA in November 1970. Due to rising costs, Weaver sold the Ju 52/3m to Cannon Aircraft in 1974 for $36,000, which only a few months later sold it to author and pilot Martin Caidin, who paid $52,500 for the old transport, and then re-registered it as N52JU. Repainted in horrendously inaccurate pseudo-Luftwaffe colours, Caidin toured the USA in the aircraft, even writing about his experiences in the book *The Saga of Iron Annie*, published in 1979.

In 1984, Caidin sold the Ju 52/3m to Lufthansa. The German airline had one condition—the aircraft had to be delivered to its maintenance facility in Hamburg. Instead of dismantling the aeroplane and transporting it across the Atlantic on a container ship, the Ju 52/3m was flown across the Atlantic. The ferry flight began on 12 December 1984, with the crew consisting of Clark Woodard, John Wilson, and Terry Ritter. The legend 'Nach Hause Going home!' was painted on the fuselage. In spite of some technical issues, the Ju 52/3m arrived in Hamburg on 28 December. Following engine shut-down, Woodard opened the cockpit window and played the German national anthem on a trumpet.

Following its triumphant arrival back on German soil, the work on refurbishing the Ju 52/3m began in earnest. The airframe was disassembled, during which it was discovered that the aeroplane

The Lufthansa Ju 52/3m 'D-AQUI' seen during the roll-out ceremony in 1986. (*Swedish Aviation Historical Society*)

was in terrible condition, with much corrosion in evidence. The BMW 132 engines were, after some consideration, replaced by Pratt & Whitney Wasps. After fifteen months of hard work, supported by many companies and numerous individuals, the Ju 52/3m was presented to Lufthansa on 6 January 1986, which also happened to be the sixtieth anniversary of Lufthansa. Although painted as D-AQUI, it was officially registered as D-CDLH (standing for Comeback Deutsche Lufthansa). The first flight took place on 1 April, exactly sixty years to the day after the first Lufthansa revenue flight. Based at Hamburg, the Ju 52/3m has since 1986 been used to 'show the flag' of Lufthansa at air shows around Europe, carrying thousands of appreciative passengers on nostalgic air experience flights. In 2015, the Ju 52/3m suffered damage to its wing main spar. Repairs are underway, and D-AQUI will return to the air.

Technikmuseum Hugo Junkers, Dessau

W Nr 6134
CO+EI
'1Z+BY'

One Ju 52/3m g4e is on display at the *Technikmuseum Hugo Junkers*, Dessau. Originally delivered to the Luftwaffe, it was lost in April 1940 during a resupply mission to Narvik, Norway. Following recovery in 1986, the Ju 52/3m was acquired by the *Technikmuseum*. It has since been restored and put on display.

Luftwaffen Museum, Wunstorf AB

W Nr 6693
DB+RD

'D-AQUI' is a frequent visitor to air shows around Europe. It is seen here at Dala-Järna, Sweden, in August 2012. (*Mikael Forslund*)

Originally delivered to Luftwaffe in 1939, this particular Ju 52/3m was issued to KGr b z bV 102. It was one of a dozen Ju 52/3ms that landed on Lake Hartvikvannet in Northern Norway in April 1940 during a resupply mission for German forces at Narvik. Along with several other Ju 52/3ms, it sank to the bottom of the lake when the ice melted. It was recovered in the mid-1980s by the Norwegian armed forces, and moved to Gardermoen. The Ju 52/3m was transferred to the Luftwaffen Museum in 1986. It was restored to static condition, being rolled out on 15 August 1987 and placed on display.

Technik Museum Speyer

W Nr 6821
VB+UP
'D-AQUI'

Another of the Ju 52/3m recovered from Lake Hartvikvannet, Norway, that has found its way back to Germany. Following recovery in 1986, it was stored at Gardermoen until acquired by Technik Museum in 1993. It is on display at the Technik Museum Speyer marked as 'D-AQUI'.

Museum für Verkehr und Technik, Berlin

W Nr 7220
D-AZAW
CJ+KA
EC-CAK
EC-ABE

One of the Ju 52/3ms recovered from Lake Hartvikvannet can be seen at Speyer. (*Ulrich Hild*)

22-108
T.2-108
'D-2201'
'D-AZAW'

One Ju 52/3m ge with a very diverse past is preserved in Berlin. Originally delivered to Lufthansa, it was subsequently transferred to Luftwaffe. In 1943, it was used by the Spanish volunteer air unit Blue Legion on the Eastern Front. It was returned to Spain, being used by both the EdA and national airline Iberia. It was eventually acquired by the *Museum für Verkehr und Technik* in May 1965 for the symbolic sum of one DM. It was painted in the colours of Lufthansa's first Ju 52/3m, D-2201. Since then, the Ju 52/3m has been repainted as 'D-AZAW'.

Hugo Junkers Kaserne, Höhn AB

C/n 053
French AF
FAP 6320
'IZ+IK'

One former FAP AAC.1 is preserved at Hugo Junkers Kaserne, Höhn AB. After being withdrawn from FAP service, it was placed on display at Ota. Eventually acquired by the Luftwaffe, the AAC.1 was dismantled and flown to Höhn in a C.160 Transall in 1975. It was restored to static display condition, being marked as 'IZ+IK'. A Luftwaffe transport unit, LTG-63, is based at Höhn AB.

The AAC.1 at Höhn in false Luftwaffe colours and markings. (*Swedish Aviation Historical Society*)

Deutsches Museum, Munich

C/n 363
Armée de l'Air 363

On 28 March 1958, an *Armée de l'Air* AAC.1 was presented to the Deutsches Museum. It is on display in *Armée de l'Air* markings, with sections of the fuselage and wing corrugated skin being removed to show the construction of the aircraft. Interestingly, this was the first Ju 52/3m/CASA 352/AAC.1 to be selected for preservation.

Flugausstellung Museum, Hermeskeil

T.2B-237
'D-CIAD'

As with several of the other CASA 352s, this particular example was obtained by Air Classik. Following delivery on 18 February 1981, it was placed on display at Dusseldorf, painted as 'D-CIAD'. In August 1982, the CASA 352 was acquired by Flugausstellung Museum, based at Hermeskeil.

Auto und Technikmuseum, Sinsheim

No less than three CASA 352s can be found at Sinsheim.

C/n 31
T.2-140

AAC.1 Toucan s/n 363 has been on public display since 1958. (*Swedish Aviation Historical Society*)

T.2B-140
N9012P
'IZ+EK'
'RJ+NP'

After being withdrawn from use, this CASA 352 was sold to the Chino, California-based Military Aircraft Restoration Corporation (MARC). It was ferried from Cuatro Vientos, arriving at Biggin Hill on 27 August 1983, having been purchased by the Junkers Ju 52/3m Flight at Rochester. Painted in Luftwaffe colours as 'IZ+EK', the CASA 352 was operated on the air show circuit until 1985. Sold to the *Auto und Technikmuseum*, it is today statically displayed at Sinsheim, painted in Luftwaffe colours as 'RJ+NP'.

C/n 100
T.2-209
T.2B-209
'D-AQUI'
'D-2527'

The hundredth-production CASA 352, this particular aeroplane was displayed at the El Avion restaurant, Plasencia, Spain. In 1988, it was sold to the *Auto und Technikmuseum*. It has been painted to represent a Lufthansa Ju 52/3m, 'D-AQUI', later being repainted as'D-2527'.

C/n 148
T.2B-257
D-CIAL

Another Sinsheim CASA 352 is located outside, being painted as 'D-2527' of Lufthansa. (*Mikael Olrog*)

D-ADAM
'D-AQUI'

After being withdrawn from active service, this CASA 352 was acquired by the German company Air Classik. It was registered as D-CIAL, and delivered to Stuttgart on 7 April 1975, being placed on display at the airport, carrying the registration D-ADAM. It was eventually removed on 2 September 1985, being consigned to storage at Malmsheim airfield. The following year, it was obtained by the *Auto und Technikmuseum*. The third CASA 352 at Sinsheim is on indoor display, hanging from the ceiling. It is painted in the colours of D-AQUI, an original Dessau-built Ju 52/3m currently airworthy with the Lufthansa *Traditionsflug*.

Flughafen München, Munich

C/n 35
T.2-144
T.2B-144
N88927
'D-CIAS'
'D-ANOY'

One CASA 352 is currently on display at Munich International Airport. After being sold off by the Spanish Air Force to the German company Air Classik, it arrived in Stuttgart, Germany, in April 1974. The aeroplane was placed on static display at Frankfurt airport on 14 September 1974. After Air Classik ceased operations, the CASA 352 was acquired by *Flughafen München* (Munich International Airport). It is on display at the airport's visitor's park, being painted as 'D-ANOY'. The original D-ANOY was used in 1937 on a route proving flight to try to establish regular passenger air traffic between Germany and China.

Great Britain

Although several CASA 352s and one AAC.1 were purchased by British aviation museums and enthusiasts during the 1970s, only one can still be found on the British Isles. The CASA 352s obtained by Doug Arnold have long since found new owners. In 2012, the IWM disposed of its AAC.1—originally donated by the Portuguese air arm in 1972–claiming that it was a 'post-war derivative' of the Ju 52/3m with numerous detail airframe differences, and that 'if it had been a genuine Junkers then it would undoubtedly be a prized part of our collection, representing a key Axis aircraft. However this particular aircraft never served in a conflict in which the UK took part, nor was it operated by enemy forces against British or Commonwealth forces'. The AAC.1 left Duxford in May 2013, having eagerly been acquired by the Polish Aviation Museum in Krakow.

The third CASA 352 at Sinsheim is painted to represent 'D-AQUI' of Lufthansa. (*Mikael Olrog*)

RAF Museum, Cosford

C/n 163
T.2-272
T.2B-272
'G-AFAP'

One CASA 352 is on display at the RAF Museum, Cosford, marked as 'G-AFAP'. Originally built in 1954, it was taken on charge by the *Ejército del Aire* on 31 December 1954 with the serial number T.2-272 (c/n 163) and code 76-8. On 25 June 1958, it entered a general overhaul at Getafe, with the BMW 132 engines being exchanged for Spanish-built ENMASA E9-C-29-750 Beta engines, the serial being amended to T.2B-272. It was retired on 2 December 1972, being consigned to storage at Cuatro Vientos. The last flight in Spanish Air Force service took place in June 1976.

Almost exactly one year later, on 3 June 1977, the CASA 352 was purchased at auction by the RAF Museum. It was flown to Biggin Hill in May 1978, with the total flight time being four hours and fifty minutes. Before departure from Cuatro Vientos, the CASA 352 had been painted in Luftwaffe colours, although the EdA serial number was retained. Interestingly, a fake RAF serial was applied. The old transport was then placed on static display at the Biggin Hill International Airfair. On 22 May 1978, it was flown to RAF Cosford for preservation. In October 1985, the Luftwaffe colours were replaced, with the CASA 352 being repainted (with British Airways providing sponsorship) to represent Ju 52/3m G-AFAP, which had been operated by the pre-war British Airways Ltd. It remains on display at Cosford.

Greece

Hellenic Air Force Museum, Dhekelia

W Nr 7607
4V+BT

On 3 October 2003, a Ju 52/3m was recovered off the island of Leros from a depth of 42 metres. The Ju 52/3m, coded 4V+BT of II./TGr 2, had been shot down by British anti-aircraft fire on 12 November 1943 during the fighting for Leros. The mortal remains of one German paratrooper were found aboard the aircraft. During the intervening years, the tail section had been badly damaged by ship's anchor, but the airframe was in quite good condition. Restoration was initiated at Dhekelia air base outside of Athens, with parts from other Ju 52/3m wrecks, donated from people of Crete and Dodecanese, being incorporated. However, due to lack of funds, completion of restoration work has been halted.

Netherlands

Aviodrome, Lelystad

C/n 166
T.2B-275
G-BFHF
'D-ADAM'
'1Z+IK'

In early 2007, a CASA 352 was obtained by the Aviodrome from Seattle-based Vulcan Warbirds. Following retirement from the *Ejército del Aire*, the CASA 352 was one of five purchased by British Warbird collector Doug Arnold. After a string of owners, the Aviodrome secured ownership of the aircraft, and it arrived at Lelystad on 23 March 2007. It has since been restored to represent a Luftwaffe Ju 52/3m, coded '1Z+IK'. The original 1Z+IK was one of several Ju 52/3ms, which, on 10 May 1940, landed on a beach near the Hague, the objective being to stop the Dutch Royal Family and members of government from escaping. Following completion of restoration, the 'Ju 52/3m' can be seen dropping German paratroopers in a diorama showing scenes from the German invasion of the Netherlands.

Norway

Three Ju 52/3ms can be found in Norway, two of which were recovered from Lake Hartvikvannet during the 1980s. Both of these are held by the Forsvarets Flysamling. At least one relatively complete Ju 52/3m still rests at the bottom of Lake Hartvikvannet. The third Ju 52/3m is a former Portuguese aircraft on display at Bodø.

Ju 52/3m 4V+BT seen soon after recovery. (*Georgios Psarras*)

Formerly owned by Doug Arnold, this CASA 352 has been painted in Luftwaffe colours as 1Z+IK, and placed in a diorama depicting the German invasion of the Netherlands. Note that the right wing is missing. (*Andrew Bird*)

Norsk Luftfartsmuseum, Bodø

W Nr 5664
AM 106
FAP 6306
'LN-DAF'

This particular Ju 52/3m was obtained in an exchange deal with the Museo do Ar. It has been restored as 'LN-KAF' of D.N.L. The original LN-KAF plied the airways along the Norwegian coast until the mid-1950s.

Forsvarets flysamling Gardermoen

W Nr 6657
CA+JY

This particular Ju 52/3m was one of eleven aircraft that landed at Lake Hartvikvannet in May 1940, just south of Narvik in northern Norway. The one-way mission was to deliver reinforcements to German forces that were occupying the city. The operation failed and the aircraft along with its cargo of troops and material were stranded on the lake. The lake was attacked by both Norwegian

A former Portuguese example, this has been restored in DNL markings as 'LN-DAF'. (*Mats Averkvist*)

and British aircraft. When the ice melted, the Ju 52/3ms sank to the bottom of the lake. Recovered in 1982, the Ju 52/3m is one of the most original Luftwaffe Ju 52/3ms to have survived. It is currently on display at *Forsvarets Flysamling Gardermoen.*

W Nr 6791
CO+EI

The second ex-Lake Hartvikvannet Ju 52/3m held by the *Forsvarets Flysamling* was originally destined for display at Bjerkvik, near Narvik, in a purpose-built building. However, this appears to have foundered, with the eventual future of the Ju 52/3m being undecided.

Poland

Muzeum Lotnictwa Polskiego, Krakow

C/n 048
Armée de l'Air
FAP 6316

In 2012, AAC.1 s/n 6316 was disposed of by the IWM to the Polish Aviation Museum, Krakow. The aeroplane arrived in Krakow in May 2013, subsequently being repainted as a Ju 52/3m of 1./KGr zbV 9. This particular unit operated on the Eastern Front in 1942. The aircraft has since been placed on outdoor display.

Portugal

A number of Ju 52/3ms and one AAC.1 still survive in Portugal, all being on charge with the *Museu do Ar*. One aircraft has been restored, with others being stored for possible use in exchange deals that would bring 'missing' aeroplanes to *Museo do Ar*. Brief details of individual aircraft are given below.

Museo do Ar, Sintra

W Nr 5655
AM 103
FAP 6303

After being withdrawn from use, this Ju 52/3m was moved to a children's playground at Évora. It has since been recovered by the *Museu do Ar*, and it is currently in outside storage in poor condition.

W Nr 5661
AM 104
FAP 6304

Following a number of years in storage, this Ju 52/3m was restored to static condition at Bodø, Norway. It was finished in night bomber colours. Although it has since returned to Portugal, the aeroplane remains in storage due to lack of space.

W Nr 501196
Lw PN+BN
NoAF YA-C
AM 201
FAP 6301

One of a pair of Ju 52/3ms obtained from the Norwegian Air Force in 1950, this remains in outside storage in poor condition, with the front fuselage being badly damaged.

W Nr 501219
Lw PR+WK
AM 200
FAP 6300
NoAF YA-B

The second former Norwegian Air Force Ju 52/3m is also kept in outside storage.

C/n 205
Armée de l'Air
FAP 6311

The above AAC.1 is kept in storage by the *Museu do Ar*. After being withdrawn from service, it was held by the Viseu Aero Club until transferred to Sintra.

Russia

W Nr 7335
Lw BV+OJ
CCCP-Л54

One Ju 52/3m recovered from its Siberian crash site is stored at Novosibirsk in anticipation of a restoration to static condition. The aircraft had crashed near Chita in eastern Siberia on 3 February 1949 while on a flight from Chita to Krasnoyarsk. According to the manufacturer's plate, the Ju 52/3m g7e was built at Bernburg in October 1941, being delivered to 1./KüFlGr.706 as BV+OJ. Subsequently transferred to KG z b V 500, the Ju 52/3m was damaged beyond local repair during a Russian air raid against Pitomnik on 5 December 1942. Following the capture of Pitomnik, the Ju 52/3m was repaired and eventually registered as CCCP-Л54 on 16 February 1944. Recovery occured during the autumn of 2003 (and continuing) with the wreck being located at Novosibirsk-Evsino Airport until 2009 at least.

Serbia

Serbian Aeronautical Museum, Belgrade

C/n 222
F-BBYB
Armée de l'Air?
JRV 7208

One former JRV AAC.1 is preserved at Surčin, Belgrade. After being withdrawn from use, the Toucan was in outside storage at Belgrade airport between 1969 and 1980. Despite being on outdoor display for a number of years, the aeroplane appears to be in good condition.

South Africa

C/n 164
T.2-164
T.2B-164
G-BFHE
ZS-UYU (marked as 'ZS-AFA')

South African Airways Historic Flight

During the late 1970s, preparations for the fiftieth anniversary of the formation of South African Airways began in earnest. Having previously acquired a DH 104 Dove, Lockheed Lodestar, and Lockheed L1649 Starliner, a search began to locate a Junkers Ju 52/3m. Various avenues were investigated, including acquiring the pair of Ju 52/3ms still present in Mozambique, and recovering the wreck of a SAAF Ju 52/3m that had crashed near Figtree, Zimbabwe. These options fell through, as did the purchase of a Ju 52/3m or AAC.1 held in store by the Portuguese Air Force Museum. The only remaining alternative was finding a Spanish-built CASA 352. In the event, one of the CASA 352s of the Warbirds of Great Britain Collection, Blackbushe, G-BFHE, was purchased, with some assistance from Airbus Industries.

On 12 May 1981, the CASA 352 left Blackbushe, being flown to Lemwerder, Germany, where it was crated for the journey southwards. On 6 July, it was offloaded at Durban and taken by road to Jan Smuts Airport where restoration was initiated. Although registered as ZS-UYU, the CASA 352 was finished as ZS-AFA, commemorating the first of SAA's Ju 52/3ms. The first flight took place on 1 February 1984, with a display slot during the fiftieth anniversary celebrations at Rand Airport. At the time of writing, the CASA 352 was awaiting renewal of its Certificate of Airworthiness.

The engineers that kept the South African CASA 352 flying. (*John-Austin Williams JAWS)*)

Spain

Four CASA 352s still survive in Spain, although none are airworthy. Brief details of the individual aircraft are given below.

Alcantarilla AB, Murcia

c/n 72
T.2-181
T.2B-181

This A-1 variant was delivered to the EdA in October 1950. It was the final EdA CASA 352 to be struck off charge, on 4 October 1978, having last served with the *Escuela Militar de Paracaidistas* (Parachute Training School). It had logged an impressive 3,116.35 hours in service.

Torrejón AB, Madrid

C/n 137
T.2B-246
792-20

Preserved at Alcantarilla AB, s/n T.2B-181 remained in EdA service for twenty-eight years. (*Luis Gonzalez Pavón*)

A model 'C' Flying Classroom variant, T.2B-246 was delivered in June 1953. It was struck off charge on 8 February 1974 with a total flying time of 1,267.45 hours. The aircraft has been given the code 792-20. Having suffered damage to its fin and rudder, the aeroplane has been moved to Getafe for restoration.

Museo del Aire, Cuatro Vientos, Madrid

Two CASA 352s are kept by the Museo del Aire, as detailed below.

C/n 102
T.2B-211
911-16

An A-3 variant, it was delivered to EdA on 21 June 1951, being assigned to the *Primer Grupo de Transporte*, based at de Alcalá de Henares, serving with this unit until June 1955. Subsequently transferred to the *Escuadrilla del Cuartel General de la Región Aérea de Levante*, Valencia. Further units included Nos 4 and 5 *Alas*. On 18 October 1963, the CASA 352 underwent an IRAN (Inspect and Repair as Necessary) overhaul. It was then transferred to the *Escuadrilla del Cuartel General de la Región Aérea Pirenaica*, Zaragoza. The final unit to operate this particular CASA 352 was 91 *Grupo*. It was struck off charge on 6 February 1973, having accumulated 1,873.15 hours in active service. The CASA 352 was transferred to the Museo del Aire on 17 May 1973.

CASA 352 s/n T.2B-246 is preserved at Torrejón AB. It is currently undergoing repairs at Getafe. (*Luis Gonzalez Pavón*)

One of two CASA 352s preserved at Cuatro Vientos, s/n T.2B-211 was transferred to the museum in 1973. (*Luis Gonzalez Pavón*)

C/n 145
T.2B-254
721-14

An A-1 variant, the CASA 352 was originally delivered to EdA in August 1954. Initially serving with the *Primer Grupo de Transporte*, the aircraft was later used operationally during the Ifni campaign by *Ala Mixta* No. 36. Before being struck off charge on 2 December 1972, the CASA 352 was used by 721 *Escuadron of the Escuela Militar de Paracaidismo*, based at Alcantarilla. It was formally transferred to the Museo del Aire on 17 May 1973.

Sweden

Parts of an early production Ju 52/3m and a CASA 352 are preserved in Sweden.

Arlanda Flygsamlingar, Arlanda

W Nr 4017
SE-ADA (ntu)
SE-ADK
SE-ADR

The cockpit section and a small wing section from SE-ADR (W Nr 4017) is preserved by the Arlanda Civil Aircraft Collection. Registered as SE-ADR on 15 July 1932, it was later purchased by ABA (Swedish Airlines). During the Second World War, it was periodically leased by *Flygvapnet*. After the end of the war, the Ju 52/3m was bought by Albin Ahrenberg, who used it for joy rides. Thousands of Swedes made their first flight in SE-ADR.

Following the last flight on 21 January 1954, the Ju 52/3m was sold to Å Liljeqvist on 19 January 1955. It was apparently offered to the National Museum of Science and Technology for preservation, but they declined due to lack of space. The Ju 52/3m remained parked at Bromma airport, eventually being purchased by the Swedish CAA to be used in firefighting exercises. However, plans for a future civil aviation museum had been formulated already in 1953 by Charlie Enger, an employee of the CAA, with the intention being to preserve a fairly large section of the fuselage of SE-ADR. However, in December 1959, only the cockpit and a small section of the left wing were cut from the Ju 52/3m. The decision about how much to preserve was decided by the size of the indoor storage area available. Sadly, the rest of the airframe succumbed in a firefighting exercise in 1961.

Svedinos Bil- och Flygmuseum, Ugglarp

C/n 33
T.2B-142
N9012N
'DP+FJ'

The cockpit section of SE-ADR survives with the Arlanda Civil Aviation Collection. (*Author's Collection*)

One CASA 352 is preserved in the privately owned *Svedino Bil- och Flygmuseum* at Ugglarp, where it is painted in Luftwaffe colours. The acquisition of the CASA 352 was part of a complicated deal during the late 1980s involving several different vintage aeroplanes for the purpose of securing a DH 98 Mosquito for *Flygvapenmuseum* (Swedish Air Force Museum). The CASA 352 had been used by the EdA until the early 1970s as serial T.2B-142, eventually being bought by the American Warbird collector Ed Tallichet, the owner of Military Aircraft Restoration Company, MARC. It was ferried to Britain by Jeff Hawke, arriving on 12 September 1979.

Registered as N9012N, it was based at Biggin Hill and briefly operated on the air show circuit. On 19 November 1982, the CASA 352 left Biggin Hill on the first leg of the flight to the USA. During the stopover in Dublin, Ireland, on 25 November, severe engine problems were encountered. The CASA 352 was subsequently damaged when high winds blew it into another aircraft. Tallichet then traded the damaged CASA 352 to *Flygvapenmuseum*, also paying for the dismantling and transportation of the aeroplane to Sweden. Following arrival in Sweden, it was rebuilt at Wing F 14 Halmstad, with the fuselage being transported to Ugglarp in May 1990. The wings were finished a year later, with the CASA being assembled and put on outdoor display. Following the construction of a hangar in the early 2000s, the CASA 352, along with several other large aircraft, was moved indoors. It has subsequently been restored as a late-production Ju 52/3mg14, being marked as DP+FJ of *Fliegerstaffel des Führers*. This particular Ju 52/3m arrived in Sweden on 2 May 1945, having earlier been at the disposal of one of Hitler's physicians, Dr Brand.

Although visitors are not allowed to enter the aircraft, a replica cockpit section of a Ju 52/3m g14e built by Bengt Hermansson was completed in 2014 and placed on display.

Restored as Ju 52/3m DP+FJ of *Fliegerstaffel des Führers*, this CASA 352 resides at the privately owned *Svedinos Bil-och Flygmuseum* at Ugglarp. (*Bengt Hermansson*)

A Ju 52/3m gl4e replica cockpit section is on display at the *Svedinos Bil-och Flygmuseum*. It is seen here under construction. (*Bengt Hermansson*)

Switzerland

Three Ju 52/3ms and one CASA 352 are currently flying in Switzerland. When the Swiss Air Force retired its trio of Ju 52/3ms in 1982, all three were eventually transferred to enthusiast societies that managed to keep the old transports flying. Today, all four are operated by the company Ju Air, which is based at Dübendorf, Switzerland. The Ju 52/3ms are used for pleasure flying, air show appearances, and movie work, each of the aircraft averaging about 200 flying hours annually. Amazingly, BMW engines are still fitted to the Ju 52/3ms. The aircraft of Ju Air are operated with the assistance of volunteers, as well as corporate sponsorship. One current sponsor is the flight luggage manufacturer Rimowa. The exterior of the Rimowa flight luggage (the luggage with the grooves) is designed in a similar fashion to the corrugated aluminium plating of the Ju 52/3m. In 2012, one of the Ju 52/3ms, HB-HOT, was flown across the Atlantic, touring Canada and the US, with stops at Oshkosh, Los Angeles, and New York. A rundown of the four aircraft is as follows:

W Nr 6580
A-701

Seen at Mönchengladbach, Germany, on 21 June 2015, Ju 52/3m HB-HOT is operated for joy flights across central Europe. On this day, the author was privileged to fly from Essen to Mönchengladbach in this particular Ju 52/3m. (*Author's Collection*)

HB-HOS

Originally delivered to the *Schweizerische Flugwaffe* in October 1939, the serial A-701 was assigned. It was used as a flying lecture hall, being fitted out to train navigators and observers. On 13 September 1948, the Ju 52/3m was registered as HB-HOS. With Swiss military aircraft being forbidden to fly outside of Switzerland, A-701 civilian registered for supply flights, carrying personnel and spares to and from Germany and Great Britain in connection with purchases of fighter aircraft. The civilian registration was cancelled on 3 September 1959. Formally withdrawn from use in 1982, the Ju 52/3m was re-registered as HB-HOS in July 1985 for the *Bundesamt für Militärflugplatz* (BAMF). A company was formed to operate the Ju 52/3m on a commercial basis, including air experience flights, air show appearances, and for movies. On 29 May 1987, the Ju 52/3m was badly damaged in a landing accident at Winningen, Germany. It was rebuilt at Altenrhein, making its first post-restoration flight on 6 October 1988. HB-HOS/A-701 remains based at Dübendorf, Switzerland.

W Nr 6595
A-702
HB-HOT

Following delivery in 1939, the Ju 52/3m was assigned the serial A-701. On 27 March 1951, it was registered as HB-HOT in connection with flights to Britain and Germany. The registration was cancelled on 3 September 1959. In 1972, the Ju 52/3m became a movie star, being painted in Luftwaffe winter camouflage for the Hollywood movie *Where Eagles Dare*. The camouflage was retained after the end of the filming, although the Luftwaffe insignia was replaced by Swiss Air Force markings. Operated by *Flugwaffe* until 1982, the Ju 52/3m was re-registered in July 1985 as HB-HOT for the *Bundesamt für Militärflugplatz* (BAMF). Operated by Ju Air, HB-HOT is based at Dübendorf.

W Nr 6610
A-703
HB-HOP

The last of the three *Flugwaffe* Ju 52/3ms was assigned the serial A-703, and, on 9 May 1947, it was registered as HB-HOP. For nearly eighteen months, it was the only Ju 52/3m used to carry personnel and spares to and from Britain and Germany. As with HB-HOS and HB-HOT, HB-HOP was cancelled on 3 September 1959. After being withdrawn from service, it was re-registered on 15 Oct 1982 as HB-HOP for *Bundesamt für Militärflugplatz* (BAMF). As with her two sisters, HB-HOP is based at Dübendorf.

T.2-165
T.2B-165
'D-CIAK'
'D-ADAM'
HB-HOY

The only suitable support vehicle for Ju 52/3m operations, a Citroën H Van. (*Author's Collection*)

In 1991, a CASA 352 was donated to Ju Air. Originally delivered to EdA as a CASA 352A-1, it was struck off charge in 1974. Obtained by the German company Air Classik, the CASA 352 delivered to Bonn-Wahn (also known as Cologne-Bonn airport), arriving on 26 April 1975. Painted as 'D-CIAK', it was placed on outdoor display at Düsseldorf airport on 19 May 1976. It was eventually repainted as 'D-ADAM' before being sold in 1987 to the *Ikarus Flugverkehrsmuseum* at Marl. The CASA 352 eventually found its way to Bochum, where it was briefly put on display. By the early 1990s, it was offered for sale, eventually being donated to the Swiss company Ju Air. It was restored to airworthy condition, the work lasting six and a half years, before flying again on 12 August 1997. It is registered as HB-HOY, being seen at air shows across Europe. It is normally based at Mönchengladbach, Germany.

Uruguay

Museo Aeronáutico, Montevideo

W Nr 5877

CX-ABA

It has been reported that the fuselage of CX-ABA is preserved at the Museo Aeronáutico in Montevideo, although positive proof of this is lacking. Nevertheless, photographic evidence of CX-ABA in outdoor storage at an unknown location in 1961 exists. Amazingly, photos taken in 1961 of the second CAUSA Ju 52/3m, CX-ABB, show the aeroplane in a dismantled state and in outdoor storage.

USA

Although the Canadian Airways Ju 52 ce CF-ARM was demonstrated in the USA in 1932, no American airlines ordered the type. Although several Ju 52/3ms were captured by American forces during the Second World War, none were brought to the USA. It was not the early 1970s that one Ju 52/3m and one CASA 352 arrived on US soil. The former had been acquired from Ecuador by a private individual, with the latter being donated to the USAF Museum by the Spanish government. The Ju 52/3m was eventually sold to Lufthansa in 1984, and is currently airworthy with the Lufthansa *Traditionsflug*. Since then, a small number of CASA 352s has been acquired by US aviation museums and enthusiasts.

National Air & Space Museum

C/n 146
T.2B-255
G-BFHD
'D-ADLH'

One CASA 352 is on display in the Boeing Aviation Hangar at the massive Udvar-Hazy Center,

The fuselage of a former CAUSA Ju 52/3m allegedly survives in Uruguay. CX-ABA survived as late as April 1962 when this picture was taken. (*Michael Magnusson*)

CX-ABB survived until August 1961 at least. (*Michael Magnusson*)

Chantilly, VA. After being retired by the Spanish Air Force, it was acquired by British Warbird collector Doug Arnold, to whom the CASA was registered as G-BFHD. It was flown to Blackbushe on 4 July 1978, and subsequently painted in Luftwaffe colours before being acquired by Brian Woodford. It was moved into storage at Bournemouth in September 1985. On 25 June 1987, the CASA 352 was obtained by the company Aces High. The aeroplane was then purchased by Lufthansa, painted as 'D-ADLH', and donated to the NASM in 1987.

National Museum of the United States Air Force

C/n 135
T.2B-244

In 1971, the Spanish government presented a CASA 352 to the USAF Museum. It was placed on public display in its original EdA colours, with the code 901-20. The Spanish colours were eventually replaced by a generic Luftwaffe camouflage scheme. The CASA 352 is currently in storage. No definite plans for the aeroplane have been formulated.

Fantasy of Flight. Florida, USA

C/n 153
T.2B-262
G-BFHG

This CASA 352 was donated by Lufthansa to the NASM in 1987. (Mikael Olrog)

Donated by the Spanish government in 1971, it was the first CASA 352 to be preserved in the USA. (*NMUSAF*)

'D2+600'
'VK+AZ'
'D-TABX'
'VZ+NK'

One of the former Doug Arnold CASA 352s, G-BFHG, is currently owned by Kermit Week's Fantasy of Flight. After being purchased by Doug Arnold in 1977, the CASA 352 was sold to movie company Aces High Ltd in November 1984. Restored to airworthiness, the CASA 352 was based at North Weald, being used in a number of movies, being painted as 'D2+600', 'VK+AZ', 'D-TABX', and 'VZ+NK'. Eventually consigned to storage, the aeroplane was sold to vintage aircraft collector Kermit Weeks in October 1992. The dismantled CASA 352 left North Weald in September 1994, arriving at Port Everglades, Florida, the following month. The CASA 352 remains in storage with Kermit Weeks Fantasy of Flight collection.

Military Aviation Museum, Virginia Beach, VA, USA

C/n 67
T.2B-176
N99069
N352JU
'1Z+AR'

One CASA 352 is kept in airworthy condition by the Military Aviation Museum, Virginia beach, Virginia, which is owned by Gerald Yagen. The CASA 352 was previously operated by the Confederate Air Force (CAF) in Luftwaffe colours, having been acquired from Spain in 1977. It was flown to the USA, leaving Biggin Hill on 5 September 1979. Upon arrival at the CAF main base at Harlingen, Texas, it was painted in Luftwaffe colours as '1Z+AR'. It was subsequently re-registered as N352JU. By the mid-2000s, the CASA 352 was sold to Gerald Yagen.

APPENDIX I

Lufthansa Ju 52/3ms

Registration	W Nr	Name	Date of Registry	Remarks
D-2201	4013	*Oswald Boelcke*	05.1932	Mid-air collision with Udet U-12 D-1296 08.1932. Repaired & Re-registered D-ADOM To DVL 34, then to RLM *Flugbereitschaft*. Cancelled 01.1939
D-2202	4015	*Richthofen*	09.1932	Re-registered D-ADYL, to RLM *Flugbereitschaft* 1934
D-2468	4019	*Joachim von Schröder*	05.1933	Re-registered D-AFIR. To *Flugbereitschaft*
D-2490	4020	*Gustav Dörr*	08.1933	Re-registered D-AFYS. To Lw, lost 1941
D-2526	4023	*Zephyr*, renamed *Emil Schäfer*	08.1933	Re-registered D-AGAV W/o London 26.11.1937
D-2527	4022	*Manfred von Richthofen*	07.1933	Re-registered D-AGUK. Later used by Göring, aircraft renamed *Kurt Wolff*. To Lw, lost 1942
D-2588	4025	*Rudolf Kleine*	08.1933	Re-registered D-AHIH. To Lw, lost 1941
D-2600(I)	4021	*Immelmann*	06.1933	Re-registered D-AHUT. To RLM, renamed *H. J. Buddecke*. Used by Hitler
D-2600(II)	4023	*Immelmann II*	04.1937	Re-registered
D-2600(III)	5702			
D-2624	4026	*Rudolf Berthold*	09.1933	Re-registered D-AJAN, leased by DDL. Transferred to FLEI Service, to Lw, lost 1941
D-2640	4028	*Werner Voss*	09.1933	Re-registered D-AZEV. To DVL as testbed, based Rechlin. To Lw
D-2649	4028	*Hermann Göring*	09.1933	W/o Frankfurt-Main
D-2650	4029	*Fritz Rumey*	10.1933	Re-registered D-AKOK. To Lw, lost 1942

D-2725	4030	*Paul Bäumer*	10.1933	Damaged London 1934, repaired. To Lw, lost 1942
D-3012	4033		02.1934	Re-registered D-AFYS, later D-AHIP? Transferred to DVS, later to DVL. To Lw, TK+HB
D-3049	4035	*Hans Gontermann*	03.1934	Re-registered D-ALAS. To Askania-Kurssteuerung. To Lw
D-3050	4036	*Joachim Schröder*, renamed *Kurt Wintgens*	1934	Re-registered D-ALUN. To Lw, lost 1941
D-3051	4037	*Kurt Wustoff*	02.1934	Re-registered D-AMAM, to Lw
D-3123	4039	*Gustav Leffers*	02.1934	Re-registered D-ANAL. To Lw, DK+QY, lost 1941
D-3127	4040	*Otto Parschau*	03.1934	Re-registered D-APAR. W/o Chanteny, France, 02.1938. Also rep as W Nr 5526
D-3131	4041	*Werner Voss*		Re-registered D-ARAM
D-3136	4042	*Paul Bilik*	03.1934	Re-registered D-ASEN. Sold to Syndicato Condor, R PP-CAY
D-3156	4043	*Kurt Wolff*	1934	Re-registered D-ABIS. Sold to Syndicato Condor, R PP-CAX, R LV-ZBD, FAA s/n T-159
D-3382	4044	*Emil Thuy*	1934	Re-registered D-ABAN. Sold to China
D-ABAQ	4066	*Manfred von Richthofen*	1935	Used by Göring as tour aircraft. To Lw, GC+AE. W/o Salsk 01.1943
D-ABES	5026	*Fritz Roth*, renamed *Herrmann Thomsen*	09.1934	To Lw, lost 1942
D-ABEW	6432	*Rudolf von Thuna*		To Lw, lost
D-ABFA	6385	*Otto Parschau*	05.1939	To Lw, TT+LN. Lost 1941
D-ABIK	4069	*Manfred von Richthofen*	1935	Used by Göring as touring aircraft. To Eurasia 03.1938
D-ABIZ	4068	*Erich Albrecht*	1935	Cancelled 05.1938
D-ABUR	5777	*Charles Haar*	06.1937	W/o on approach to Frankfurt 04.01.1938
D-ABVF	5954	*Franz Wagner*	04.1938	To Lw, lost 1942
D-ABYF	4045	*E Schofer*	09.1934	W/o Hallgarten 25.04.1935. Possibly D-AJYR?
D-ACAW	6700	*Heinrich Mathy*		
D-ACBE	6550	*Emil Schofer*		To Lw, lost
D-ACBO	4059	*V von Neubrand*		P/i ZS-AFB, to Iberia R M-CABO
D-ABCQ	7439 *			To Lw CK+QO, to Aero O/Y R OH-LAP, ret DLH re-registered D-ACDA, to Lw VK+PJ destroyed 1945

D-ACEP	6386	*Adolf von Tutscheck*		To Lw, lost
D-ADAL	4046	*Karl Allmenroder*	05.1934	To Deruluft, leased to Lw
D-ADBO	6387	*Otto von Beaulieu-Marconay*		To Lw, lost
D-ADBW	6650	*Emil Thuy*	1939	To Lw, lost
D-ADED	6046	*Viktor Neubrand*	08.1938	Leased to Lw
D-ADEF	4070	*Adolf Schirmer*	1935	To RLM and then to Lw. Lost 03.02.1944
D-ADEK	5278	*Anton Schulz*	1935	Leased to Lw
D-ADER	5120	*Hans Wende*	1935	Sold to Syndicato Condor R PP-CBE
D-ADHF	6066	*W Hohndorf*	1939	To Lw, lost 1943
D-ADQU	640605	*Karl Noack*	1943	To Lw, W/o Norway 10.1944
D-ADQV	640608	*Hermann Stache*	1943	W/o Hestnuten, Norway 16.10.1944.
D-ADQW	640610	*Harry Rother*	1943	W/o Belgrade 01.1944
D-AEAC	7493			Leased from Aero O/Y. Re-registered D-AXVA, OH-LAO, D-AEAC, CK+QR, SE-BUD, VH-BUV, VH-GSS
D-AENF	5283	*DLH Sudamerika Aconcagua*		Lufthansa Südamerika. Later to Ecuador, and then to USAAF
D-AFAM	5673		01.1937	DLH, used by Goebbels. To lw, DD+MD
D-AFCD	5938	DLH		
D-AFER	4047	*Franz Buchner*	04.1934	Leased to Lw
D-AFFQ	6057	*G. Falke*		To Lw
D-AFOP	5800	*Karl Hochmut*		Crashed on take-off, Hannover, 30.08.1939
D-AGAK	5685	*Ulrich Neckel*	03.1937	Leased by Lw
D-AGBI	6659	*Max von Mulzer*	1939	To Lw, destroyed 1942
D-AGDA	4080	*Weige von Froreich*		P/i OE-LAM. To Lw, destroyed 24.02.1942
D-AGEI	5472	*Karl Allmenröder*		To Eurasia, R EU XIX
D-AGES	5104	*Otto Kissenberth*		To Eurasia, R EU XVII
D-AGFD	4060	*Otto Parschau*		P/i ZS-AFC, to Iberia, R M-CABU, EC-AAI
D-AGIQ	5272	*Martin Zander*		To OA-HHC
D-AGIS	4048	*Kormoran*		To Deruluft
D-AGOB	6452	*H Handke*		To Lw, destroyed in 1942
D-AGOO	5555	*Fritz Simon*		Leased to Kriegsmarine
D-AGST	5261	*Maipo*		Lufthansa Sudamerika. R PP-CAZ
D-AGTC	6030	*Wilhelm Cuno*		To Lw, 7U+LM. W/o 02.1943
D-AHAL	5034	*Otto Bernert*		Leased by Lw, destroyed in 1944
D-AHFN	6047	*H Kircheldorf*		To Lw, destroyed in 1942
D-AHGA	6775	*Paul Bilik*		Leased by Lw

D-AHGB	6800	*Rudolf Kleine*		R PP-CBP
D-AHMS	6042	*Martin Zander*		To Lw, destroyed in 1941
D-AHUS	4049	*Heinrich Kroll*		To Deruluft
D-AIAG	7244	*Olaf Bielenstein*		To Lw, destroyed in 1944
D-AIAH	7268	*Karl Hochmuth*		To Lw, destroyed in 1944
D-AIAO	4064			P/i I-BIZI, MM60410, Lw AI+AO
D-AIAT	6765			P/i I-BOAN, MM60413
D-AJAO	5478	*Robert Weinhard*		R PP-CBD
D-AJAT	4076			P/i OE-LAK, to Iberia, R M-CABC, EC-AAK
D-AJIM	4050	*Ulrich Neckel*		Later renamed *H. Göring*. Crashed Frankfurt 19.09.1936
D-AJYR	4045	*Emil Schäfer*		Crashed Hallgarten, 25.04.1935
D-AKEQ	5590	*Gustav Rubrius*		To ÖLAG, R OE-LAN, RR D-AKEQ. To Lw as SG+BM. Destroyed in 1942
D-AKIQ	5429	*William Langanke*		To DNL, R LN-DAF
D-AKUO	5484	*Paul Bilik*		To Lw, destroyed in 1941
D-AKYS	5098	*Emil Thuy*		To Iberia, R M-CABY
D-ALAM	5740	*William Langanke*		To Lw, destroyed in 1943
D-ALAN	5010	*Eduard Dostler*		To Lw, destroyed in 1942
D-ALUE	5502	*Joachim von Schröder*		To Eurasia, R EU XX
D-ALUG	5093	*Josef Zauritz*		To Lw as TU+BP
D-ALYL	5180	*Hans Loeb*		Renamed *XI Olympiade* in 1936. To ÖLAG, R OE-LAR
D-AMAK	5294	*Volkmar von Arnim*		To Eurasia, R EU XVIII
D-AMAQ	5109	*Max von Mulzer*		To Syndicato Condor, R PP-CBH
D-AMEI	5734	*Fritz von Roeth*	03.1937	To Lw, destroyed in 1943
D-AMFR	5933	*Ludwig Hautzmayer*		To ÖLAG, R OE-LAS. Ret DLH. Wreck still at Tempelhof in 1947
D-AMIP	4072	*Fritz Erb*		To Eurasia, R EU XXIV
D-AMIQ	5272	*Martin Zander*		R OA-HHC
D-AMIT	5060	*O von Beaulieu-Marconnay*	10.1934	
D-AMYE	5656	*DLH Sudamerika Anden*		To Syndicato Condor, R PP-CBL
D-AMYM	6015			To Iberia, R EC-DAM
D-ANAZ	5128	*Willi Charlett*		To Lw
D-ANEN	5072	*Fritz Puetter*		
D-ANJH	5947	*Hans Loeb*	03.1938	W/o Mingaladon, Burma, 03.08.1938
D-ANOL	5014	*Albert Dosselbach*	08.1934	To Lw, destroyed in 1941
D-ANOP	4077	*Fritz Simon*		To DNL, R LN-DAE
D-ANOY	5663	*Rudolf von Thuna*		W/o Vienna, 12.1938

D-ANXG	5979	*Hans Kirschstein*	07.1938	
D-ANYF	4071	*Erich Pust*		To Lw, destroyed in 1941
D-ANYK	5329	*Wilhelm Schmidt*	09.1934	To Eurasia, R EU XV
D-AOCA	640996	*Harry Rother*		S/d over Balkans, 17.04.1944
D-APAA	5606	*Otto Kissenberth*	10.1936	Operated on behalf of German government
D-APAJ	7029	*Erich Pust*		Leased Lw
D-APDF	5915	*DLH Sudamerika Ecuador*		To HC-SAB
D-APEF	4075	*Karl Wessel*		To Syndicato Condor, R PP-CBG
D-APGU	6734	*Bruno Rodschinka*		Leased Lw
D-APOK	5074	*Max von Mulzer*		To SEDTA, R HC-
D-APOO	5506	*Heinrich Kroll*	07.1936	Crashed into mountains near Tabarz, 11.1936
D-APOR	4079	*Olaf Bielenstein*		To Syndicato Condor, R PP-CBF
D-APUP	5682	*Marschall von Bieberstein*		Leased Lw
D-APXD	6149	*Robert Untucht*		To Lw
D-APZX	6750	*Raoul Stoiavlevic*		To Lw, surrendered at Oslo, Norway. To RAF, VM908.
D-AQAR	4055	*Walter Hohndorf*		To Syndicato Condor, R PP-
D-AQAU	5650			
D-AQIJ	7053			To Iberia, R EC-CAJ. W/o Echterding, 09.1944
D-AQUB	5751			Captured in Norway 09.04.1940. P/i LN-DAI. W/o Hommelvik 03.1943
D-AQUI	5489	*Fritz Simon*		To DNL, R LN-DAH
D-AQUQ	5053	*Adolf von Tutschek*		To SEDTA, R HC-SAC
D-ARAD	6171	*Volkmar von Arnim*		Surrendered at Oslo, Norway, 05.1945
D-ARCK	6779	*Van Vloten*		Leased Lw
D-ARDS	5919	*Robert Weichard*	03.1938	To Lw, destroyed in 1942
D-AREB	6442	*Charles Haar*	06.1939	Leased Lw, destroyed in 1942
D-AREN	4051	*Phoenix*	06.1934	To Deruluft. W/o near Stettin, 31.01.1935
D-ARIW	6180	*Joachim Blankenburg*		To Lw, destroyed in 1941
D-ARUN	4038	*DLH Sudamerika Caicara*		To Syndicato Condor, R PP-CAV
D-ARVU	6790	*Hans Wende*		Leased Lw
D-ARYS	5043	*Hans Kirschstein*		To OA-HHB
D-ASDI	7077	*Rudolf Kleine*		To Lw, destroyed in 1944
D-ASEV	5087			To Eurasia, R EU XXI
D-ASFD	6014	*Heinrich Mathy*	08.1938	To Eurasia, R EU XXV
D-ASHE	?	*Friedrich Dahmen*		W/o Konorn, Hungary

D-ASHY	7089	*Gustav Doerr*		Leased Lw
D-ASIH	5078	*Ruolf Windisch*		W/o Grand Bornand, 31.12.1936
D-ASIS	4074	*Wilhelm Cuno*		To Eurasia, R EU XXIII
D-ASLG	6369	*Alfred Viereck*		To Lw
D-ASPI	6710			P/i I-BIOS
D-ASUI	5498	*Hans Berr*		W/o Lauf-Pregnintz, 11.1936
D-ATAK	5169	*Marschall von Bieberstein*		W/o Hannover, 28.11.1936. Scrapped
D-ATAO	5748	*Alfred Bauer*		Leased Lw. Damaged in landing accident at Banak, Norway, 26.09.1942
D-ATAW	7160	*Lothar von Richthofen*		Leased Lw
D-ATDB	5940	*Walter Bayer*	03.1938	Leased Lw
D-ATEA	5727			P/i OE-LAP. Leased Lw
D-ATON	4054	*Erwin Böhme*	09.1934	
D-ATYZ	5797	*Hans Hackmack*	08.1937	Leased Lw
D-AUJA	5851	*Otto Fink*		To Iberia, R M-CABE
D-AUJG	5942	*Hans Wende*		P/i I-ABJZ. W/o Llaveria, Spain, 04.08.1939
D-AUKE	5854	*Willie Rabe*		To Iberia, R M-CABA
D-AUSS	641382	*Josef Langheld*		W/o Prujavor, Bosnia, 10 01.1945
D-AUWA	?	*Gerhard Amann*		S/d Ivanca, 02.09.1944
D-AUXZ	7172	*Otto von Beaulieu-Marconnay*	1941	W/o Gabrenze, Bulgaria, 20.10.1941
D-AVAJ	6370	*Olaf Bielenstein*		To Lw, destroyed
D-AWAS	6561	*Joachim Blankenburg*		Lost over Mediterranean, 21.02.1944
D-AXAT	5693	*Rudolf Windisch*	02.1937	W/o Munich, 12.1937
D-AXES	4052	*Hans Berr*		To Deruluft, To Lw, destroyed 1941
D-AXFH	6372	*H E Lochner*		To Lw, destroyed
D-AXFL	7196			To Iberia, R EC-CAK
D-AXOS	5023	*Oswald Boelcke*	08.1934	Leased Lw
D-AXUT	5022	*Lothar von Richthofen*		To Iberia, R M-CABO
D-AYGX	7208	*Johannes Horoldt*		W/o Bucharest, 10.1942
D-AYHO	5860	*Peter Strasser*	02.1938	
D-AZAN	5021	*Joachim Schröder*	08.1934	To Lw
D-AZAW	7220			To Iberia, R EC-CAN
D-AZIR	1301	*Fritz Erb*	12.1937	
D-AZIS	5020	*Von Bülow, Horst Wessel*	09.1934	To Lw 1942

APPENDIX II

Luftwaffe Ju 52/3m Transport Units, 1939–1943

Unit	Formed	Code	Remarks
KGr z b V 1			
Stab/KG z b V 1	26.098.1939 at Stendal	1Z	Redesignated Stab/TG 1, 05.1943
I./KG z b V 1	26.08.1939 at Gardelegen	1Z	Redesignated I./TG 1, 05.1943
II./KG z b V 1	26.08.1939 at Stendal	1Z	Redesignated II./TG 1, 05.1943
III./KG z b V 1	26.08.1939 at Stendal	1Z	Redesignated III./TG 1, 05.1943
IV./KGr z b V 1	26.08.1939 at Liegnitz	1Z	Redesignated IV.TG 1, 05.1943
KGr z b V 2		G6	Redesignated I./TG 2, 05.1943
Stab/KG z b V 2	26.08.1939 at Küpper-Sagan	G6	Redesignated Stab/TG 3, 05.1943
I./KG z b V 2	26.08.1939 at Sorau	G6	Redesignated KGr z b V 12, 11.1939
II./KG z b V 2	26.08.1939 at Freiwaldau	G6	Disbanded 11.1939
III./KG z b V 2	26.08.1939 at Freiwaldau	G6	Disbanded 11.1939
IV./KG z b V 2	26.08.1939 at Breslau-Gandau	G6	Disbanded 11.1939
KGr z b V 3			Redesignated Stab/TG 4, 05.1943
KGr z b V 4	01.08.1938 at Tutow		Disbanded 22.10.1938. Reformed 02.1942 in Germany, but disbanded 15.06.42
KGr z b V 5	01.08.1938 at Fassberg	L5, S3	Disbanded 22.10.1938. Reformed 18.08.1942 at Radom with He 111 transports. Became TGr 30 in 05.1943
KGr z b V 6	01.08.1938 at Lechfeld		Disbanded 22.10.1938. Reformed 02.1942 equipped with Ju 52/3ms and Ju 86s
KGr z b V 7	02.1942 in Germany		Equipped with Ju 52/3ms and Ju 86s. Disbanded, 03.1942 and absorbed by KGr z b V 4, 5, and 8
KGr z b V 8	02.1942 in Germany		Disbanded, 03.1942
KGr z b V 9	26.08.1939 at Eichwalde	4V	Redesignated I./TG 3, 05.1943
KGr z b V 10	11.1939 at Hagenow*		Redesignated IV./KG z b V 1 in 03.1940
KGr z b V 11	04.1940 at Lippstadt		Disbanded 01.06.1940
KGr z b V 12	04.1940 at Störmede		Formed from I./KG z b V 2. Disbanded 01.06.1940

KGr z b V 40	01.1941 in Berlin	9P	Disbanded 06.1941 due to heavy losses sustained during Operation Mercury
KGr z b V 50	01.1941 in Berlin	9P	Redesignated II./TG 3, 05.1943
KGr z b V 60	01.1941 in Berlin	9P	Disbanded 06.1941 due to heavy losses sustained during Operation Mercury
KGr z b V 101	03.1940 in Germany	G6	Disbanded 06.1940 apart from Stab/ KGr z b V 101. Reformed 02.1941. Disbanded 06.1941 due to heavy losses sustained during Operation Mercury
KGr z b V 102	03.1940 at Oldenburg	G6	Disbanded 05.05.1940 apart from Stab/ KGr z b V 102. Reformed 02.1941. Redesignated III./TG 3, 05.1943
KGr z b V 103	03.1940 at Schleswig	G6	Disbanded 05.05.1940
KGr z b V 104	03.1940 at Stade	G6	Disbanded 06.1940 apart from Stab/ KGr z b V 104. Reformed 02.1941. Redesignated II./TG 5, 05.1943
KGr z b V 105	03.1940 at Kiel-Holtenau	G6	Disbanded 06.1940, apart from Stab/ KGr z b V 105. Reformed 02.1941. Redesignated I./TG 4, 05.1943
KGr z b V 106	03.1940 at Uetersehen		Disbanded 06.1940, apart from Stab/ KGr z b V 106. Reformed in 02.1941 at Insterburg. Became III./TG 2, 05.1943
KGr z b V 107	03.1940 at Fuhlsbüttel		Disbanded 06.1940, most aircraft transferred to KGr z b V 108
KGr z b V 108	03.1940 at Rantum 6 Hörnum	6I	Redesignated TGr 20, 05.1943
Stab/KGr z b V 172	26.08.1939 at Berlin-Tempelhof		Disbanded 10.11.1939
I./KGr z b V 172	26.08.1939 at Berlin-Tempelhof	4V	Redesignated 03.1940 as III./KG z b V 1, and then to III./KGr z b V 172 later the same month. Redesignated IV./TG 3, 05.1943
II./KGr z b V 172	26.08.1939 at Berlin-Tempelhof	4V	Disbanded 11.1939. Reformed 02.1941 in Berlin. Disbanded 06.1941. Redesignated Blindflugschule 6
III./KGr z b V 172	26.08.1939 at Berlin-Tempelhof		Redesignated 03.1940 as I./KGr z b V 172.
KGr z b V 300	11.1941 in Germany		Operated in training role from 01.07.1942. Redesignated Ergänzungsgruppe, 05.1943
KGr z b V 400	10.12.1941 at Munchen-Riem	4F	Redesignated III./TG 4, 05.1943
KGr z b V 500	10.12.1941 at Munchen-Riem	J6	Known as KGr z b V Oels between 01.1942 and 04.1942. Redesignated I./ TG 4, 05.1943
KGr z b V 600	17.12.1941 in East Prussia	C6	Redesignated I./TG 2, 05.1943

KGr z b V 700	17.12.1941 in East Prussia	7V	Redesignated IV./TG 4, 05.1943. Exchanged Ju 52/3ms for LeO 451s in 03.1943
KGr z b V 800	17.12.1941 in East Prussia		Redesignated II./TG 2, 05.1943
KGr z b V 900	17.12.1941 in East Prussia		Began to receive Me 323s in 02.1943. Redesignated III./TG 5 05.1943. Not completed
KGr z b V 999	17.12.1941 in East Prussia		Disbanded, 04.1942
KGr z b V Brindisi (S-13)			Disbanded, 12.1942
KGr z b V Frankfurt			Disbanded, 02.1943
KGr z b V Neapel (S-7)			Disbanded, 01.1943. Aircraft to KGR z b V 400
KGr z b V Posen			Disbanded, 04.1942
KGr z b V Reggio (S-11)			Disbanded, 12.1942
KGr z b V Wittstock			Disbanded, 02.1943

APPENDIX III

Luftwaffe Ju 52/3m Transport Units, May 1943–February 1945

Unit	Formed	Code	Remarks
Stab/TG 1	15.05.1943		From Stab/KGr z b V 1. Disbanded 30.09.1944
I./TG 1	15.05.1943		From I./KGr z b V 1. 3 and 4./TG 1 disbanded 30.09.1944
II./TG 1	15.05.1943		From II./KGr z b V 1. Disbanded 04.09.1944
III./TG 1	15.05.1943		From III./KGr z b V 1. Disbanded 04.09.1944
IV./TG 1	15.05.1943		From IV./KGr z b V 1. Disbanded 30.01.1945
Stab/TG 2	15.05.1943		From Stab/KGr z b V 3. Disbanded 30.09.1944
I./TG 2	15.05.1943		From KGr z b V 600. Disbanded 04.09.1944
II./TG 2	15.05.1943		From KGr z b V 800.
III./TG 2	15.05.1943		From KGr z b V 106
Stab/TG 3	15.05.1943		From Stab/KG z b V 2
I./TG 3	15.05.1943		No information
II./TG 3	15.05.1943		No information
III./TG 3	15.05.1943		No information
IV./TG 3	15.05.1943		No information
Stab/TG 4	15.05.1943	G6	Disbanded 20.10.1944
I./TG 4	15.05.1943	G6	From KGr z b V 105. Disbanded 10.1944
II./TG 4	15.05.1943	G6	From KGr z b V 500. Disbanded 10.1944
III./TG 4	15.05.1943	G6	From KGr z b V 400.
TGr 20	15.05.1943 in Oslo, Norway	7U	1./TGr 20 redesignated in 11.1943 as *Seetransportstaffel* 3. In its place, a new 1./TGr 20 was formed. 4./TGr 20 disbanded 10.1944

APPENDIX IV
AAC.1 Production List

C/n	Delivered to	Registration	Date registered	Remarks
1	Air France	F-BAJA	08.12.1944	Wfu 01.01.1948. TT 3,086 hrs in Air France service. To *Armée de l'Air* 21.01.1948 SoC 13.01.1961
2	Air France	F-BAJB	13.12.1944	W/o Pau, France 04.10.1947. TT 2,689 hrs
3	Air France	F-BAJC	15.12.1944	Op by Transporte Aérien Militaire (TAM), III/15 in 04.1945, serial 03, later 003. Ret Air France. Wfu 01.07.1947. TT 2,745 hrs. To AIA 31.07.1947 for *Armée de l'Air*
4	Air France	F-BAJD	18.12.1944	Del. to TAM. Ret to French gvt 05.08.1948. R F-ZIFA. Accident 30.09.1947, repaired. Del. to IGN 05.08.1948, used until 1952 at least. Derelict at Creil by mid-1950s
5	Air France	F-BAJE	03.01.1945	Del. to TAM, III/15 in 1945, then to Air France. Wfu 01.09.1947. TT 2,811 hrs. To AIA 02.09.1947 for del. to *Armée de l'Air* 05.1948. Transferred to FAP 19.12.1960
6	Air France	F-BAJG	03.01.1945	Based in Madagascar, to Europe 01.04.47. Wfu 01.02.48. TT 2,972 hrs. To AIA for del. to *Armée de l'Air* 23.03.1948. SoC 08.03.1959
7	Air France	F-BAJH	17.01.1945	Based in Madagascar, to Europe 01.04.1947. Wfu 01.01.1948. TT 3,048 hrs. To AIA 05.05.1948. Re-registered F-BBOF 18.03.1954 for Préfecture de Police. Take-off accident at Gonesse 17.03.1955. Cancelled 05.1955. Derelict at Le Bourget
8	Air France	F-BAJI	22.01.1945	Del. to TAM. Named Lt A. Texier. Based Dakar 01.04.1947. Wfu 01.01.1948. TT 3,174 hrs. To AIA 03.01.1948 for del. to *Armée de l'Air*. SoC ?
9	*Armée de l'Air*	9	01.1948	SoC 15.07.1959
10	*Armée de l'Air*	10		Transferred to Aéronavale at unknown date
11	*Armée de l'Air*	11		No information
12	*Armée de l'Air*	12		No information

13	Air France	F-BAJJ	27.01.1945	Based Dakar 01.04.1947. Wfu 01.01.1948. TT 2,932 hrs. To AIA 21.01.1948 for del. to *Armée de l'Air* 01.1948. SoC 15.05.1960. NB; Acceptance date rep as 11.1954
14	Air France	F-BAJK	26.01.1945	Based in Madagascar 08.1948. Cancelled 19.12.1957, wfu in Madagascar. TT 5,704 hrs
15	Air France	F-BAJL	01.02.1945	Based in Madagascar. Cancelled 01.1953, wfu in Madagascar. TT 5,830 hrs
16	Air France	F-BAJM	06.02.1945	Named Oradour. Wfu 01.01.1948. TT 2,496 hrs. To AIA 12.03.1948 for del. to *Armée de l'Air*. Dbr 15.12.1957
17	Air France	F-BAJN	09.02.1945	Del. to TAM. Wfu 01.04.1948. TT 3,113 hrs. To AIA 22.04.1948 for del. to *Armée de l'Air*. SoC 20.10.1955
18	Air France	F-BAJO	26.01.1945	Del. to TAM. Based Dakar. Wfu 01.03.1948. TT 3,056 hrs. To AIA 05.03.1948 for del. to *Armée de l'Air*. Acceptance date noted as 07.1955. SoC 31.12.1960
19	Air France	F-BAJP	23.01.1945	Del. to TAM, III/15. Coll on ground with *Armée de l'Air* Ju 88 at Le Bourget 10.09.1945. Dbr. TT 502 hrs
20	Air France	F-BAJS	01.02.1945	Op by CEPM for Postal service. Hit high tension wires near Pau 28.06.1948. Dbr. TT 1,349 hrs
21	*Armée de l'Air*	21		No information
22	*Armée de l'Air*	22		No information
23	*Armée de l'Air*	23		No information
24				No information
25				No information
26	*Aéronavale*			Op by French West Africa Flt, based Dakar. Coded DK-1
27				No information
28	*Aéronavale*			No information
29	*Armée de l'Air*	29		No information
30	*Aéronavale*			No information
31	*Armée de l'Air* *			No information
32	*Aéronavale*			No information
33	*Aéronavale*			No information
34				No information
35				No information
36	Air France	F-BAJU	03.04.1945	Based Dakar. Wfu 08.01.1948. TT 2,985 hrs. To AIA 21.01.1948 for del. to *Armée de l'Air*. SoC 22.11.1960. NB; also rep as del. to CSA as OK-PCC, and later re-registered as OK-PDC
37	Air France	F-BAJV	03.04.1945	Wfu 01.09.1947. TT 2,490 hrs. To AIA 27.08.1947 for del. to *Armée de l'Air*. Acceptance date 01.1951. SoC 13.01.1961

38	Air France	F-BAJX	03.04.1945	Wfu 01.08.1947. TT 2,285 hrs. To AIA 19.08.1947 for del. to *Armée de l'Air*. Dbr near Turenne, 12.04.1957
39	*Aéronavale*			No information
40				No information
41	*Aéronavale*			No information
42	CEV			Transferred to *Aéronavale*
43	*Armée de l'Air*			No information
44	Air France	F-BAJT	05.04.1945	W/o on landing at Le Bourget 19.08.1946. TT 1,212 hrs
45	*Armée de l'Air*	45	03.1945	SoC 22.11.1960
46	*Armée de l'Air*			No information
47	*Armée de l'Air*			No information
48	*Armée de l'Air*		07.1948	Transferred to FAP 26.11.1960
49	*Armée de l'Air*			No information
50	*Armée de l'Air*			No information
51	*Armée de l'Air*			No information
52	*Armée de l'Air*			No information
53	*Armée de l'Air*		10.1944 [*sic.*]	Transferred to FAP 26.11.1960
54	*Armée de l'Air*			No information
55	Air France	F-BAKM	25.04.1945	W/o Freetown, Sierra Leone 20.03.1947 TT 2,052 hrs. Scrapped
56	Air France	F-BAKN	25.04.1945	Used for tests between 04.1946 and 12.1951. Wfu 01.12.1951 due to airframe corrosion. TT 3,013 hrs. To DGER for spares recovery and ground instructional training
57	Air France	F-BAKO	30.04.1945	Dbr at Mahon, Balearic Isles, 04.02.1946 TT 664 hrs
58	Air France	F-BAKP	01.05.1945	W/o at Le Bourget in landing accident due to poor visibility, 05.03.1947. TT 1,114 hrs. Scrapped
59	Air France	F-BAKQ	04.05.1945	Based at Algiers. Wfu 01.09.1947. TT 2,340 hrs. To AIA 02.09.1947 for del. to *Armée de l'Air*. SoC 26.08.1960
60	Air France	F-BAKR	01.05.1945	Based at Algiers. Wfu 01.10.1947. TT 2,235 hrs. To AIA 20.10.1947 for del. to *Armée de l'Air*
61				No information
62				No information
63	*Armée de l'Air*			SoC 13.01.1961
64	*Armée de l'Air*			No information
65	?			No information
66			May 1945	SoC 16.05.1960
67	*Armée de l'Air*			No information
68	*Armée de l'Air*			SoC 15.07.1959
69	*Armée de l'Air*			SoC 08.09.1955
70	*Armée de l'Air*	070/70		No information

71	Air France	F-BAKS		Intended for Air France, but diverted to *Armée de l'Air* Service du Materiel. Dbr 23.06.1951
72	Air France	F-BAKT	30.05.1945	Based at Algiers. Wfu 01.01.1949. TT 3,440 hrs. To AIA 12.01.1949 for del. to *Armée de l'Air*. Serial 072 rep as F-RBDA of *Armée de l'Air* in Indochina
73	Air France	F-BAKU	31.05.1945	Named Beiteddine. R LR-AAC 31.05.1945. Wfu 01.04.1948. TT 1,511 hrs. R LR-AAC in 06.1946. Sold to Yemen goverment on 13.04.1948, but not confirmed as delivered
74	Air France	F-BAKV	15.06.1945	W/o Dakar-Yoff, 07.06.1947. TT 2,264 hrs. Scrapped
75	Air France	F-BAKX	04.06.1945	Returned to *Armée de l'Air* 01.10.1946. TT 609 hrs. Loaned to Service des Telecommunications et de Signalisation. To AIA 12.02.1949 for del. to *Armée de l'Air*
76	Air France	F-BAKY	05.06.1945	Wfu 01.06.1948. TT 3,131 hrs. To AIA 23.06.1948 for del. to *Armée de l'Air*. Dbr 18.10.1955
77	Air France	F-BAKZ	06.06.1945	Wfu, and del. to Yugoslavia 15.08.1945. Possibly op by JRV until 1947, and later transferred to JAT
78	Air France	F-BALA	01.06.1945	Wfu and del. to Yugoslavia 18.08.1945. Possibly operated by JRV until 1947, and later transferred to JAT
79	Air France	F-BALB	16.06.1945	Wfu 19.02.1948. TT 2,939 hrs. To AIA 19.02.1948 for del. to *Armée de l'Air*. SoC 1953
80	Air France	F-BALC	16.06.1945	Wfu 19.08.1948. TT 3,155 hrs. To AIA 19.08.1948 for del. to *Armée de l'Air*. SoC 11.03.1959
81	*Armée de l'Air*			SoC 06.04.1960
82	*Armée de l'Air*			SoC 24.06.1955
83	*Armée de l'Air*			No information
84	*Armée de l'Air*		01.1945	SoC 08.1959
85	*Armée de l'Air*			No information
86	*Armée de l'Air*		01.1945	SoC 21.03.1962, likely last AAC.1 in *Armée de l'Air* service
87	*Armée de l'Air*		06.1945	SoC 18.01.1960
88	*Armée de l'Air*			No information
89	*Armée de l'Air*		06.1945	SoC 14.09.1955
90	Air France	F-BALE	02.07.1945	Based in Madagascar. W/o Miandrivazo, Madagascar, 10.04.1953. TT 5,135 hrs
91	Air France	F-BALF	02.07.1945	W/o Eseka, 100 km WSW Yaounde, French Cameroon, 01.07.1947. TT 2,016 hrs
92	Air France	F-BALG	19.07.1945	Based at Dakar. Wfu 01.07.1948. TT 2,422 hrs. To AIA 23.08.1948 for del. to *Armée de l'Air*. SoC 18.10.1955

93	(Air France)	(F-BALH)	Ntu	Diverted to SCRAS, later to *Armée de l'Air*. SoC 11.12.1956
94	Air France	F-BALI	04.07.1945	Based at Dakar. Wfu 25.02.1948. TT 2,412 hrs. To AIA 22.03.1948 for del. to *Armée de l'Air*
95	Air France	F-BALJ	04.07.1945	Based at Algiers. Wfu 02.12.1947. TT 2,497 hrs. To AIA 05.02.1948 for del. to *Armée de l'Air*. SoC 20.10.1955
96	Air France	F-BALK	04.07.1945	W/o Belo, Madagascar, 02.02.1946. TT 202 hrs
97	Air France	F-BALL	04.07.1945	Transferred to SCRAS
98	Air France	F-BALM	09.07.1945	Transferred to SCRAS
99	Air France	F-BALN	05.07.1945	Op by CPEM on Postal service. Wfu 08.04. 1949. TT 3,069 hrs. To AIA 21.04.1949 for del. to *Armée de l'Air*
100	Air France	F-BALO	05.07.1945	Based in Madagascar. Wfu 16.02.1951. TT 3,864 hrs. Scrapped Ivato 17.07.1951
101	Air France	F-BAKK	19.07.1945	Wfu 01.10.1948. TT 3,157 hrs. To AIA 05.11.1948 for del. to *Armée de l'Air*. SoC 12.1959
102	Air France	F-BAKL	09.07.1945	W/o in landing accident, Toulouse, 23.11.1945. TT 82 hrs. Scrapped
103	Air France	F-BALD	06.08.1945	Wfu 08.07.1948. TT 3,131 hrs. Transferred to CATRE when retained by French gvt for del. to IGN. Re-registered F-BALD 1949. Cancelled 1951. To *Armée de l'Air*
104	Air France	F-BAMO	20.07.1945	Op by CPEM on Postal service. Wfu 05.11.1948. TT 2,440 hrs. To AIA 29.11.1948 for del. to *Armée de l'Air*. SoC 15.07.1959
105	Air France	F-BAMP	27.07.1945	Wfu 01.07.1947. TT 1,802 hrs. To AIA 30.07.1947 for del. to *Armée de l'Air*. SoC 07.03.1956
106	Air France	F-BAMQ	06.08.1945	Based in Madagascar. Crashed on flight between Andapa and Sambava, 01.01.1952. TT 4,795 hrs
107	Air France	F-BAMR	25.07.1945	Based at Dakar. Wfu 10.02.1948. TT 2,526 hrs. To AIA 10.02.1948 for del. to *Armée de l'Air*
108	(Air France)	(F-BAMS)		Intended for Air France, but damaged in accident pre-delivery. Rep and del. to *Armée de l'Air*
109	Air France	F-BAMT	27.07.1945	Named *Safa*. Cancelled, and R L-RAMT 02.1946, then to LR-AAJ 16.05.1946. Wfu 01.04.1949. TT 1,559 hrs. To Syrian Air Force, 09.07.1949
110	Air France	F-BAMU	06.08.1945	Wfu 18.10.1948. TT 3,120 hrs. To AIA 26.11.1948 for del. to *Armée de l'Air*. Acceptance date noted as 'Mar 46'. SoC 12.06.1955

111	Air France	F-BAMV	28.07.1945	Ret to DTA late 1945. Sold to Bata Shoe Company, and R OK-ZBB
112	Air France	F-BAMX	31.07.1945	Based in Madagascar, later to Brazzaville. Wfu 01.11.1951. TT 4,276 hrs
113	Air France	F-BAMY	31.07.1945	Named *El Arz* (*Cedar trees*). Canc, and R L-RAMY 02.1946, then to LR-AAI 16.05.1946. Wfu 01.07.1949. TT 1,460 hrs. To Syrian Air Force, 07.07.1949
114	Air France	F-BAMZ	30.07.1945	Returned to French gvt 20.09.1945. To *Armée de l'Air* 02.1947. SoC 29.03.1960
115	Air France	F-BANA	30.07.1945	Based at Dakar. Wfu 04.02.1948. TT 2,487 hrs. To AIA 02.03.1948 for del. to *Armée de l'Air*. SoC 14.09.1960
116	*Armée de l'Air*			No information
117	*Armée de l'Air*		07.1945	SoC 06.04.1960
118	*Armée de l'Air*			Dbr 21.08.1951
119	*Armée de l'Air*			No information
120	*Armée de l'Air*		07.1945	SoC 16.05.1960
121	*Armée de l'Air*		10.1947	SoC 06.04.1960
122				No information
123	*Armée de l'Air*			W/o near Ambatondrazaka, Madagascar, 12.06.1950
124	*Armée de l'Air*			No information
125	*Armée de l'Air*		06.1948	SoC 26.07.1960
126	*Armée de l'Air* or CEV			Noted i/s 08.1953, coded H
127	*Armée de l'Air*			Transferred to FAP 10.12.1960
128	*Armée de l'Air*			SoC 26.09.1955
129	*Armée de l'Air*		1945	So 13.01.1961
130	*Armée de l'Air*		08.1945	SoC 12.1959
131	*Armée de l'Air*			SoC 26.08.1960
132				No information
133	*Armée de l'Air*		08.1945	SoC 22.11.1960
134				No information
135				No information
136	Air France	F-BANB	05.09.1945	W/o near Gémenos, France, 05.06.1947. TT 1,226 hrs. Also rep as del. to CSA, R OK-ZCA, later OK-ZDO
137	Air France	F-BANC	25.09.1945	Initially based in Madagascar, later transferred to Brazzaville. Wfu 12.09.1951
138	Air France	F-BAND	03.10.1945	Destined for Yugoslavia 18.09.1949, but diverted to Czechoslovakia 07.12.1945
139	Air France	F-BANE	03.10.1945	Destined for Yugoslavia 19.09.1945, but diverted to Czechoslovakia 07.12.1945
140	Air France	F-BANF	07.10.1945	Registered owner SGACC? To AIA 09.02.1949 for del. to *Armée de l'Air*

141	Air France	F-BANG	05.10.1945	Wfu 01.10.1946. TT 468 hrs. To Service des Télécommunications et de Signalisation, 03.1946. To SALS for Centre National de St-Yan, 11.1949. Wfu 07.10.1952
142	Air France	F-BANH	02.10.1945	Based at Dakar. Wfu 01.10.1948. TT 2,937 hrs. To AIA 12.10.1948 for del. to *Armée de l'Air*. SoC 09.09.1959
143	(Air France)	(F-BANI)		Intended for Air France, but retained by DTI. Del. to *Armée de l'Air*
144	Air France	F-BANJ	22.10.1945	Based in Madagascar. Wfu on 01.05.1953. TT 4,939 hrs. Scrapped at Tananarive
145	Air France	F-BANK	10.10.1945	Based in Madagascar. Dbr in landing accident at Antalaha, Madagascar, 05.12.1952. TT 4,145 hrs. Scrapped
146		(F-BANL)		Registration reserved, ntu
147	Air France	F-BANM	22.10.1945	Wfu 01.10.1948. TT 3,047 hrs. To AIA 14.10.1948 for del. to *Armée de l'Air*. SoC 13.01.1961
148	Air France	F-BANN	26.10.1945	To Aero Portuguesa, 10.04.1946. R CS-ADA. Ret Air France 12.1947. R F-BDYK. Based in Madagascar. Owned by SGACC 01–08.1949. Wfu in Madagascar 01.11.1954. TT 3,770 hrs
149	Air France	F-BANO	16.10.1945	W/o at Le Bourget in ground collision with *Armée de l'Air* Nord NC 900
150	Air France	F-BANP	25.10.1945	W/o Le Bouscat, near Bordeaux, 13.01.1946
151	(Air France)	F-BANQ	20.01.1947	Intended for Air France, but retained by Service du Matériel. To SOCTRA, and then to *Aigle Azur*, 24.04.1947. To Roger Colin/ CATI 09.1948. W/o in Red River, Lao Kay, 21.08.1949
152	Air France	F-BANR	20.11.1945	Wfu 01.09.1947. TT 1,847 hrs. To AIA 27.08.1947. To *Armée de l'Air* 05.1948. SoC 16.05.1960
153	Bata Shoe Company	OK-ZBC		Initially reserved as F-BANS, but ntu
154	*Aéronavale*			No information
155	*Armée de l'Air*		08.1947	SoC 24.06.1955
156				No information
157	*Armée de l'Air*			SoC 08.09.1955
158	*Armée de l'Air*			No information
159				No information
160	*Armée de l'Air*			No information
161	*Armée de l'Air*		02.1945	SoC 09.09.1959
162	*Armée de l'Air*		02.1945	SoC 09.09.1959
163	*Armée de l'Air*			No information
164	*Armée de l'Air*			No information
165	*Armée de l'Air*		10.1945	SoC 22.11.1960

166	*Armée de l’Air*			No information
167	*Armée de l’Air*		02.1948	SoC 26.07.1960
168	*Armée de l’Air*		11.1945	SoC 13.01.1961
169	*Armée de l’Air*			SoC 18.02.1955
170	*Armée de l’Air*		11.1945	SoC 26.07.1960
171	*Armée de l’Air*			No information
172	*Armée de l’Air*		12.1945	SoC 1955
173	*Armée de l’Air*			SoC 09.08.1956
174	*Armée de l’Air*			Dbr Indochina
175	*Armée de l’Air*			SoC 20.10.1955?
176	*Armée de l’Air*			No information
177	*Armée de l’Air*			No information
178	*Armée de l’Air*			SoC 03.02.1960
179	*Armée de l’Air*		01.1946	SoC 17.03.1958
180	*Armée de l’Air*		1945	R F-BDYI 22.03.1950 for ONERA. To CEV 16.02.1956. Re-registered F-BDYI 01.1957 for Société Transfrigorifique du Soudan. Cancelled 05.1959
181	*Armée de l’Air*		07.1947	SoC 16.05.1960
182	*Armée de l’Air*			No information
183				No information
184	*Armée de l’Air*			No information
185	*Armée de l’Air*			No information
186	*Armée de l’Air*		01.1946	SoC 09.09.1959
187				No information
188	*Armée de l’Air*		08.1948	SoC 16.05.1960
189	*Armée de l’Air*			No information
190	*Armée de l’Air*			SoC 17.10.1955
191	*Armée de l’Air*			No information
192	*Armée de l’Air*			Dbr 05.06.1948
193	*Armée de l’Air*			No information
194	SANA	F-BBZF	14.06.1946	To DTI 18.12.1947. To *Armée de l’Air* 03.1948
195	Air France	F-BBYJ	16.04.1946	Wfu 01.10.1946. TT 368 hrs. Transferred to SGACC. To AIR 02.02.1949 for del. to *Armée de l’Air*. SoC 01.04.1960
196	SANA	F-BBZG	14.06.1946	Canc. To *Armée de l’Air* 03.1948. SoC 22.11.1956
197	SANA	F-BBZH	01.06.1946	Cancelled 1947. To *Armée de l’Air* 1948. SoC 26.07.1960
198				No information
199	*Armée de l’Air*			No information
200	*Armée de l’Air*			No information
201	*Armée de l’Air*			No information
202	SANA	F-BBZI	01.06.1946	W/o St Auban-sur-Erguerre, 11.1946. Cancelled 1948, noted as transferred to *Armée de l’Air*, unlikely

203	SANA	F-BBZJ	14.06.1946	To DTI 05.04.1948 for del. to *Armée de l'Air*. SoC 18.03.1959
204	SANA	F-BBZK	14.06.1946	Named *Kerjean*. Cancelled 11.1949. To *Armée de l'Air* already in 03.1948. SoC 16.05.1960
205	*Armée de l'Air*		02.1947	Transferred to FAP 02.12.1960
206	*Armée de l'Air*			SoC 08.09.1955
207	*Armée de l'Air*			SoC 15.07.1959
208				No information
209	*Armée de l'Air*			No information
210	*Armée de l'Air*			No information
211				No information
212	*Armée de l'Air*			No information
213	*Armée de l'Air*		04.1946	SoC 13.01.1961
214	*Armée de l'Air*		11.1947	SoC 14.09.1960
215	*Armée de l'Air*		04.1946	SoC 22.11.1960
216	*Aéronavale*			French West Africa Flt, serial DK-2. To Escadrille 55S, code 55S.32. SoC 1961, derelict at Vélizy-Villacoublay by mid-1960s. By 1975, stored at Dugny-Le Bourget for *Musée de l'Air*. On display Le Bourget by 1978. Painted as c/n 334 of *Armée de l'Air*
217				No information
218	*Aéronavale*			No information
219	*Aéronavale*			No information
220	*Aéronavale*			No information
221	*Société Transatlantique Aérien* (STA), Lyons	F-BBYA	26.08.1946	Intended for *Armée de l'Air*, but w/o in Indochina before transfer
222	STA	F-BBYB	26.08.1946	Cancelled 1948, to French military. O/h by AIA, at Clermont-Ferrand late 1949 prior to transfer to Yugoslavia. To JRV, s/n 7208. Preserved
223	STA	F-BBYC	26.08.1946	Ditched in Mediterranean 03.03.1948 on flight between Oran and Perpignan
224	Air Ocean	F-BBYD	14.06.1946	Impounded at Le Bourget, 18.06.1947. Canc. To French gvt, and to either *Armée de l'Air* or *Aéronavale*
225	LASO France	F-BBYE	27.06.1946	To *Armée de l'Air* 11.09.1947
226	Air France	F-BBYF	20.06.1946	Wfu 01.01.1949. TT 1,433 hrs. Transferred to SGACC. To Autrex, 24.08.1950. W/o Lao Kay, 26.07.1951
227	Air France	F-BBYG	13.06.1946	Registered owner SGACC. W/o Niamey, Niger, 30.04.1947
228	Air France	F-BBYH	20.06.1946	Wfu 01.10.1946. TT 193 hrs. Transferred to SGACC. To Service des Télécommunications et de Signalisation. W/o Aoulef, Central Sahara, Algeria, 30.05.1947

229	Air France	F-BBYI	17.06.1946	Registered owner SGACC. To SALS for Centre National de Mitry Mory. Cancelled 1951
230	LASO France	F-BBZL	27.06.1946	To Air Atlas, to Autrex, 12.1949. W/o Hanoi, Vietnam, 11.05.1953
231	TAI	F-BBYK	25.06.1946	W/o Mont Ventoux, 06.01.1947
232	TAI	F-BBYL	25.06.1946	W/o Mariagne, 26.10.1946
233	TAI	F-BBYM	25.06.1946	Cancelled 1947. To *Armée de l'Air* 10.1948. SoC 04.09.1958
234	TAI	F-BBYN	25.06.1946	Cancelled 1947. To *Armée de l'Air*. Transferred to FAP 10.12.1960
235	*Avions Bleu*	F-BBYO	15.10.1946	Named Martine. To DTIA late 1948 for del. to *Armée de l'Air*
236	*Aéro Cargo*	F-BBYP	21.10.1946	Cancelled 1947. To *Armée de l'Air*, dbr 14.09.1955
237	CGT	F-BBYQ	05.07.4196	To *Armée de l'Air* (?) 10.1948
238	CGT	F-BBYR	05.07.1946	Cancelled 1948. To *Armée de l'Air*
239	CGT	F-BBYS	19.07.1946	To *Armée de l'Air* 06.1948. SoC 01.04.1960
240	Air Nolis	F-BBYT	19.07.1946	To Air Atlas. Cancelled 1951, scrapped
241		(F-BBYS)		Intended for unknown civilian operator, diverted to *Aéronavale* in 05.1946. SoC 26.07.1960
242	*Armée de l'Air*			No information
243	*Armée de l'Air*			GT 1/64. Destroyed by Việt Minh artillery at Hòa Bình, Vietnam, 11.01.1952.
244	*Armée de l'Air*			No information
245	*Armée de l'Air*			No information
246	Air Nolis	F-BBZA	27.06.1946	Occasionally op by subsidiary LASO. To SANA 1947. Named *Kerguern*. To *Armée de l'Air* 11.09.1947. SoC 13.01.1961
247	SOCOTRA	F-BBZB	10.07.1946	Ret to DTIA 02.1948
248	SOCOTRA	F-BBZC	30.07.1946	W/o Casablanca, 08.01.1948
249	CGT	F-BBZD	19.07.1946	To *Armée de l'Air* 10.1948
250	*Aigle Azur*	F-BBZE	30.09.1946	Damaged Calvi, 05.04.1947. Repaired. Cancelled 1948. To *Armée de l'Air* 05.1948. SoC 26.07.1960
251	*Armée de l'Air*			Transferred to *Aéronavale*. Served with 28F 15.10.1953
252	*Armée de l'Air*		05.1946	SoC 16.05.1960
253	*Armée de l'Air*			No information
254	*Armée de l'Air*			No information
255	*Armée de l'Air*		06.1946	Transferred to FAP 10.12.1960
256	*Armée de l'Air*			SoC 15.07.1959
257	*Armée de l'Air*			No information
258	*Armée de l'Air*		08.1946	Transferred to FAP 22.11.1960
259	*Armée de l'Air*		10.1945	SoC 22.11.1960
260				No information. Possibly to *Aéronavale*
261	*Aéronavale*			No information

262	*Aéronavale*			No information
263	*Aéronavale*			No information
264	*Aéronavale*			No information
265	*Aéronavale*			No information
266	*Armée de l'Air*		10.1945	SoC 13.01.1961
267	*Armée de l'Air*		06.1946	SoC 06.10.1958. Transferred to Aerovavale
268	*Armée de l'Air*			No information
269	*Armée de l'Air*			No information
270	*Armée de l'Air*		10.1948	SoC 16.05.1960
271	*Armée de l'Air*			No information
272	*Armée de l'Air*		06.1946	SoC 08.09.1955
273	*Armée de l'Air*		06.1946	SoC 08.09.1955
274	*Armée de l'Air*			No information
275				No information
276	*Armée de l'Air*		01.1948	SoC 26.07.1960
277	*Armée de l'Air*			
278	*Armée de l'Air*			Possibly transferred to *Aéronavale*
279	*Armée de l'Air*		09.1946	SoC 08.09.1955
280	*Armée de l'Air*			
281	Air Ocean	F-BCAA	23.07.1946	W/o Sefrou, Atlas Mtns, 10.10.1946
282	SOCOTRA	F-BCAB	30.07.1946	To DTIA 02.1948 for del. to *Armée de l'Air*
283	*Compagnie de Transports Aériens* Languedoc Roussillon	F-BCAC	26.07.1946	To *Aigle Azur* 31.01.1947. Cancelled 1947. To *Armée de l'Air*. SoC 22.11.1960
284	CTA Languedoc Roussillon	F-BCAD	29.07.1946	W/o St Leger, Haute Vienne, 01.11.1946
285	STA, ntu			Diverted to *Armée de l'Air* 27.06.1957
286	CTA Languedoc Roussillon	F-BCAE	19.07.1946	To *Armée de l'Air* 09.1948. SoC 26.08.1960
287	*Société des Transports Aériens* Alpes Provence	F-BCAF	13.03.1947	Cancelled 02.1950. To *Armée de l'Air*, accepted 07.1946. SoC 01.04.1960
288	DTIA/CATRE	F-BEPD	16.03.1948	Built 46, initially served with French military. Ret to military service in 05.1948
289	*Armée de l'Air*		12.1946	SoC 03.02.1960
290	*Armée de l'Air*		12.1946	SoC 20.10.1955
291	*Armée de l'Air*		12.1946	Transferred to FAP 02.12.1960
292	*Armée de l'Air*		09.1946	SoC 09.01.1957
293	*Armée de l'Air*			No information
294	*Armée de l'Air*			SoC 26.08.1960
295	*Armée de l'Air*		10.1945	SoC 20.10.1955
296	*Armée de l'Air*		1947	SoC 13.01.1961
297	*Armée de l'Air*			W/o Muna, Nigeria?, 14.12.1948

298	*Armée de l'Air*			W/o Atar, Mauretania (French Sahara)?, 05.12.1957
299	*Armée de l'Air*			No information
300	*Armée de l'Air*		07.1948	SoC 14.09.1960
301	*Armée de l'Air*			No information
302	*Armée de l'Air*		10.1946	SoC 29.08.1955
303	*Armée de l'Air*			No information
304	*Armée de l'Air*		10.1945	SoC 14.09.1960. Also op by CEV
305	*Armée de l'Air*			No information
306	*Armée de l'Air*			Dbr 09.07.1951
307	*Armée de l'Air*			No information
308	Air Ocean	F-BCHE	07.11.1946	Impounded at Le Bourget, 18.06.1947. Transferred to French gvt 02.1949
309	*Avions Bleu*	F-BCHF	05.12.1946	Cancelled 02.1949. To *Armée de l'Air*
310	*Aéro Cargo*	F-BCHB	18.09.1946	Cancelled 11.1947. Transferred to CEV (repaired i/s 09.10.1947), and later to *Armée de l'Air*
311	SOCOTRA	F-BCHD	18.10.1946	W/o Baherya, Egypt, 23.10.1946
312	TAI	F-BCHA	12.09.1946	Cancelled 1947. Transferred to *Armée de l'Air*
313	CGT	F-BCHC	28.09.1946	Canc. To *Armée de l'Air* 09.1948. Dbr 23.04.1958
314	*Aigle Azur*	F-BCHI	15.11.1946	Cancelled 1947. Transferred to *Armée de l'Air*
315	Air Atlas	F-BCHJ	06.11.1946	Named Ville d'Oudja. To Air France, 08.1950. Initially based at Brazzaville, and later in Madagascar. Wfu 08.1956.
316	Air Atlas	F-BCHK	05.11.1946	Named *Ville de Marrakech*. Damaged, and to *Avions Bleu* 11.1948. Del. to JAT 03.1951. R YU-ACE. W/o near Rijeka, Yugoslavia, 29.06.1951
317	Air Atlas	F-BCHL	09.12.1946	Named *Ville de Rabat*. Cancelled 1951, scrapped
318	Air Atlas	F-BCHM	09.11.1946	Cancelled 1951, scrapped
319	Air Atlas	F-BCHN	21.10.1946	Cancelled 08.1952, scrapped
320	Air Atlas	F-BCHO	06.11.1946	Named *Ville de Safi*. Cancelled 11.1949. Possibly to *Armée de l'Air*
321	Air Atlas	F-BCHP	10.12.1946	To Air France 08.1950. Initially based at Brazzaville, and later in Madagascar. Wfu 01.06.1954. TT 2,242 hrs
322	Air Atlas	F-BCHQ	26.10.1946	Named *Ville de Meknes*. W/o El Ajeb, 30.08.1947
323	*Aéro Cargo*	F-BCHH	27.10.1946	W/o Vichy, 14.07.1947
324	TAI	F-BCHG	25.10.1946	Cancelled 1947. Transferred to *Armée de l'Air*. SoC 18.09.1956
325	*Armée de l'Air*		03.1947	Transferred to FAP 19.12.1960
326	*Armée de l'Air*		10.1946	SoC 16.05.1960
327	*Armée de l'Air*			W/o Morocco, 10.10.1956

328	*Armée de l'Air*			No information
329				No information
330	*Armée de l'Air*			No information
331	*Armée de l'Air*			No information
332	*Armée de l'Air*			No information
333	*Armée de l'Air*			No information
334	*Armée de l'Air*			Op by GT 1/34. Coded DG. The Toucan preserved by the *Musée de l'Air* is painted in these markings. The former curator and director of the *Musée de l'Air*, General Lissarague, flew serial 334 in Indochina
335	*Armée de l'Air*			No information
336	*Armée de l'Air*			No information
337				No information
338				No information
339	*Armée de l'Air*			No information
340	*Armée de l'Air*			No information
341	*Armée de l'Air*		12.1946	W/o Djanet, French Sahara, 02.04.1957. NB; Crash site unconfirmed
342	*Armée de l'Air*		12.1946	SoC 13.06.1955
343	*Armée de l'Air*			
344	*Armée de l'Air*			
345	*Armée de l'Air*			
346				No information
347				No information
348	*Armée de l'Air*		12.1948	SoC 13.01.1961
349	*Armée de l'Air*			
350	*Armée de l'Air*		12.1946	SoC 23.11.1960
351	*Armée de l'Air*			SoC 20.10.1955
352	*Armée de l'Air*			
353	*Armée de l'Air*			
354	*Armée de l'Air*			SoC 13.01.1961
355	*Armée de l'Air*		03.1947	SoC 09.09.1959
356	*Armée de l'Air*		03.1947	SoC 08.12.1960
357	*Armée de l'Air*		10.1945	Transferred to FAP 19.12.1960
358	*Armée de l'Air*		03.1947	SoC 16.05.1960
359	*Armée de l'Air*			SoC 20.10.1955
360	*Armée de l'Air*			
361	*Armée de l'Air*		07.1947	SoC 13.01.1961
362	*Armée de l'Air*		04.1947	SoC 08.09.1955
363	*Armée de l'Air*			SoC and transferred to Deutsches Museum, Munich, Germany, 28.03.1958. Preserved in *Armée de l'Air* markings
364	*Armée de l'Air*			SoC 02.02.1955
365	*Armée de l'Air*		04.1947	SoC 26.07.1960
366	*Armée de l'Air*		04.1947	SoC 26.07.1960

367	*Armée de l'Air*		1947	
368	*Armée de l'Air*			
369	*Armée de l'Air*			
370	*Armée de l'Air*			
371	*Armée de l'Air*			
372	*Armée de l'Air*		10.1945?	SoC 16.05.1960
373	*Armée de l'Air*			
374	*Armée de l'Air*			
375	*Armée de l'Air*		06.1947	SoC 16.05.1960
376	*Armée de l'Air*		06.1947	Transferred to *Aéronavale* 06.10.1958
377	*Armée de l'Air*			
378	*Armée de l'Air*			
379	*Armée de l'Air*			
380	SOCOTRA	F-BCHT	04.07.1947	To DTIA 02.1948, then to *Armée de l'Air.** Transferred to *Aéronavale*
381	TAI	F-BCHX	30.05.1947	Cancelled 1947. To *Armée de l'Air*. SoC 20.10.1955
382	*Aigle Azur*	F-BDYA	09.07.1947	Cancelled 1948. To *Armée de l'Air*
383	*Avions Bleu*	F-BCHZ	17.06.1947	Cancelled 1947. To *Armée de l'Air*
384	STA Alpes Provence	F-BCHR	23.07.1947	Cancelled 06.1948. To *Armée de l'Air* or *Aéronavale*
385	*Avions Bleu*	F-BDYB	08.07.1947	Cancelled 02.1949. To *Armée de l'Air*
386	SOCOTRA	F-BCHU	14.08.1947	Ret to DTIA 02.1948
387	TAI	F-BCHY	09.07.1947	To *Aigle Azur*. Cancelled 12.1948. To *Armée de l'Air*
388	*Avions Bleu*	F-BDYJ	26.09.1947	Cancelled 01.1949. To *Armée de l'Air*. Transferred to *Aéronavale*, i/s with 28F 52/53
389	*Armée de l'Air*		Built 07.1947	Loaned to Préfecture de Police 51, R F-BDYM 18.04.1951. Cancelled 05.1955, and ret to *Armée de l'Air*
390	LASO France	F-BDYD	14.08.1947	Cancelled 02.1950. To *Armée de l'Air*, acceptance date 08.1947. SoC 26.07.1960
391	SOCOTRA	F-BCHV	17.11.1947	Ret to DTIA 02.1948. To *Armée de l'Air* SoC 18.01.1960
392	*Armée de l'Air*			Transferred to FAP 19.12.1960
393	*Armée de l'Air*		1947	SoC 22.11.1960
394				No information
395				No information
396				No information
397	*Armée de l'Air*			SoC 18.01.1960
398	GECAT	F-BDYF	20.08.1947	Cancelled 02.1950. To *Armée de l'Air*. Dbr 29.06.1957
399	GECAT	F-BDYG	16.09.1947	Cancelled 02.1950
400	SOCOTRA	F-BDYC	17.11.1947	Cancelled 02.1950
401	SANA	F-BDYH	10.1947	W/o near St Claude, Jura, France, 13.11.1947

402				No information
403	*Armée de l'Air*		03.1948	SoC 13.01.1961
404	SANA	F-BDYE	06.08.1947	To DTI, then leased to Air Fret Transimex in late 1949. W/o Rovigo, Algeria, 12.05.1951
405	*Armée de l'Air*			
406	*Armée de l'Air*			
407	*Armée de l'Air*			
408	*Armée de l'Air*		03.1948	SoC 20.10.1955
409	*Armée de l'Air*		09.1947	Transferred to *Aéronavale* 06.10.1958
410	*Armée de l'Air*			
411	*Armée de l'Air*			
412	*Armée de l'Air*			
413	*Armée de l'Air*			
414	*Armée de l'Air*			
415	*Armée de l'Air*		12.1948	SoC 22.11.1960

APPENDIX V

CASA 352 Production List

C/n	EdA Serial	Variant	ToC	SoC	Remarks
1	T.2-110, T.2B-110	A-1, cv A-3	07.1944	02.06.1959	
2	T.2-111, T.2B-111	A-3	01.1945	27.06.1968	
3	T.2-112, T.2B-112	A-1	01.1945	21.10.1972	
4	T.2-113, T.2B-113	?	07.1945		
5	T.2-114	?	04.1945	15.04.1952	W/o Morón 26.06.1951
6	T.2-115, T.2B-115	C	04.1946	14.07.1971	
7	T.2-116, T.2B-116	C	05.1947	16.01.1960	
8	T.2-117	?	10.1947		
9	T.2-118, T.2B-118	A-3	06.1947	31.07.1971	Last flight 03.02.1970
10	T.2-119, T.2B-119	C	06.1947	22.10.1965	
11	T.2-120	?	07.1947		
12	T.2-121	?	10.1947	18.06.1954	
13	T.2-122	?	10.1947	12.07.1957	
14	T.2-123, T.2B-123	C	02.1948	21.10.1959	Destroyed in forced landing 16.07.1959
15	T.2-124	A-1	02.1948	29.09.1950	W/o Galapagar, Madrid, 29.09.1950. Five killed
16	T.2-125, T.2B-125	A-1	10.1948	04.03.1970	
17	T.2-126, T.2B-126	C	04.1948	09.11.1965	
18	T.2-127, T.2B-127	A-3 cv A-1	03.1948		Preserved at Hermeskeil, Germany
19	T.2-128, T.2B-128	A-1	06.1948	24.07.1965	
20	T.2-129, T.2B-129	A-1	07.1948	17.11.1961	
21	T.2-130, T.2B-130	A-1	10.1948	26.07.1973	W/o
22	T.2-131, T.2B-131	C	10.1948	23.01.1970	
23	T.2-132, T.2B-132	C	08.1948	20.02.1970	
24	T.2-133, T.2B-133	C	10.1948	08.11.1969	
25	T.2-134	?	10.1948		
26	T.2-135, T.2B-135	A-1 cv A-3	12.1948	08.06.1973	Tested by Grupo EV with Sirio engine
27	T.2-136, T.2B-136	A-1	12.1948	23.06.1959	W/o, hit ground following engine failure
28	T.2-137, T.2B-137	?	10.1949	30.11.1955	
29	T.2-138	?	01.1949	29.02.1952	
30	T.2-139, T.2B-139	C	12.1948	29.09.1959	W/o

31	T.2-140, T.2B-140	A-3 cv A-1	12.1948	26.06.1976	Preserved at Sinsheim, Germany
32	T.2-141	?	03.1949	18.04.1951	W/o
33	T.2-142, T.2B-142	A-3	03.1949	27.06.1975	Preserved in Sweden
34	T.2-143	A-3	03.1949	18.01.1950	W/o Tobarra, Albacete, 18.01.1950. Fourteen killed
35	T.2-144, T.2B-144	C cv A-1	06.1949		Preserved in Munich, Germany
36	T.2-145, T.2B-145	?	06.1949		Damaged at Carmoli, 18.01.1954
37	T.2-146, T.2B-146	A-3	06.1949	31.07.1971	
38	T.2-147, T.2B-147	?	06.1949	25.02.1956	W/o Rascafrias, Madrid, 20.12.1955. Four killed
39	T.2-148, T.2B-148	A-3	06.1949	26.11.1975	Preserved in Canada
40	T.2-149, T.2B-149	A-1	06.1949	21.10.1972	
41	T.2-150, T.2B-150	A-1	08.1949	09.03.1960	W/o
42	T.2-151, T.2B-151	A-1	08.1949	24.06.1970	
43	T.2-152, T.2B-152	?	08.1949	04.10.1955	
44	T.2-153, T.2B-153	A-3	07.1949	09.09.1969	
45	T.2-154, T.2B-154	A-1	10.1949	09.09.1969	
46	T.2-155, T.2B-155	A-1	11.1949	09.04.1961	W/o
47	T.2-156, T.2B-156	A-1 cv A-3	10.1949	28.08.1970	W/o
48	T.2-157, T.2B-157	C	11.1949	22.02.1971	W/o
49	T.2-158, T.2B-158	A-3	11.1949	06.07.1968	W/o
50	T.2-159, T.2B-159	C	03.1950	20.02.1969	
51	T.2-160	?	03.1950	18.04.1951	W/o Tobarra, Albacete
52	T.2-161, T.2B-161	A-3	12.1949	20.02.1969	
53	T.2-162, T.2B-162	?	03.1950	29.08.1954	W/o
54	T.2-163, T.2B-163	A-3	03.1950	14.07.1971	
55	T.2-164, T.2B-164	C	03.1950	09.01.1969	Currently airworthy in South Africa as ZS-UYU
56	T.2-165, T.2B-165	A-1	04.1950	30.11.1974	Currently airworthy in Switzerland as HB-HOY
57	T.2-166, T.2B-166	C	03.1950	12.03.1969	Preserved in the Netherlands
58	T.2-167	A-3	03.1950		SoC after being damaged in forced landing
59	T.2-168, T.2B-168	A-1	04.1950	06.01.1968	
60	T.2-169	?	05.1950		Damaged in forced landing at Cella, Teruel, 21.10.1951
61	T.2-170, T.2B-170	A-1	05.1950	11.02.1969	
62	T.2-171, T.2B-171	A-1	06.1950	12.08.1959	
63	T.2-172, T.2B-172	A-1 cv A-3	06.1950	25.07.1968	
64	T.2-173, T.2B-173	C	07.1950	23.12.1974	
65	T.2-174, T.2B-174	A-3	07.1950	17.11.1970	
66	T.2-175	?	07.1950	27.06.1952	SoC following accident
67	T.2-176, T.2B-176	C	07.1950	08.06.1965	Currently airworthy in USA as N352JU

68	T.2-177, T.2B-177	C	08.1950	26.07.1973	
69	T.2-178, T.2B-178	A-1	08.1950	08.02.1962	
70	T.2-179, T.2B-179	A-1	08.1950	12.03.1969	
71	T.2-180, T.2B-180	A-1	10.1950	12.03.1969	
72	T.2-181, T.2B-181	A-1	10.1950	04.10.1978	Preserved at Alcantarilla AB, Spain
73	T.2-182	?	10.1950	12.11.1951	Destroyed at los Llanos, Albacete
74	T.2-183, T.2B-183	A-1	10.1950	03.03.1966	W/o on landing Lanzarote
75	T.2-184	A-1	12.1950	05.01.1955	
76	T.2-185	?	12.1950	17.03.1953	W/o on take-off from Getafe
77	T.2-186, T.2B-186	A-3	12.1950	02.03.1968	
78	T.2-187, T.2B-187	C	12.1950	20.02.1969	
79	T.2-188, T.2B-188	A-3 cv A-1	12.1950	14.07.1971	Collision on ground with T.2B-262 at el Aaiún
80	T.2-189, T.2B-189	A-1	12.1950	09.10.1972	
81	T.2-190, T.2B-190	C cv A-1	12.1950	18.04.1974	
82	T.2-191, T.2B-191	C	12.1950	08.10.1969	W/o 26.06.1969 at Alicante
83	T.2-192, T.2B-192	C	12.1950	23.09.1961	
84	T.2-193, T.2B-193	C cv A-1	12.1950	08.06.1973	
85	T.2-194, T.2B-194	A-1	01.1951	13.03.1969	Damaged in forced landing on 15.02.1965
86	T.2-195, T.2B-195	A-1 cv C cv A-1	03.1951	02.08.1968	
87	T.2-196, T.2B-196	A-1	03.1951	12.03.1969	
88	T.2-197	A-1	03.1951	26.02.1955	
89	T.2-198, T.2B-198	A-1	03.1951	14.07.1971	Coll on ground 27.07.71 with T.2B-264 at Gando
90	T.2-199, T.2B-199	A-1	04.1951	30.04.1960	SoC due to corrosion
91	T.2-200	A-1	05.1951	06.11.1957	
92	T.2-201, T.2B-201	A-1	05.1951	05.02.1968	
93	T.2-202, T.2B-202	A-3	04.1951	30.06.1974	
94	T.2-203, T.2B-203	A-3	04.1951	05.01.1968	
95	T.2-204, T.2B-204	A-3	05.1951	01.02.1961	W/o 19.02.1960 at Menorca
96	T.2-205, T.2B-205	A-3 cv A-1	05.1951	28.08.1970	
97	T.2-206, T.2B-206	A-3	05.1951		
98	T.2-207, T.2B-207	A-3	06.1951	10.07.1957	
99	T.2-208, T.2B-208	A-3	05.1951	21.01.1959	Ditched in sea 03.07.1957, 7 km off Tétouan. One killed
100	T.2-209, T.2B-209	A-3	05.1951	21.09.1972	Preserved at Sinsheim, Germany
101	T.2-210, T.2B-210	A-3 cv A-1	06.1951	31.07.1971	
102	T.2-211, T.2B-211	A-3	06.1951	06.02.1973	Preserved at Museo del Aire, Cuatro Vientos, Spain
103	T.2-212, T.2B-212	A-3 cv A-1	07.1951	28.11.1975	R EC-ZRI. Currently airworthy in France as F-AZJU

104	T.2-213, T.2B-213	A-3 cv C	07.1951	05.02.1968	
105	T.2-214, T.2B-214	A-1	07.1951	10.06.1969	
106	T.2-215, T.2B-215	A-1	08.1951	13.09.1956	W/o 01.02.1956 at Penãlba de la Segovia. Seven killed
107	T.2-216, T.2B-216	A-1	09.1951	01.12.1958	W/o 05.10.1956 at Ojos Negros, Teruel. Five killed
108	T.2-217, T.2B-217	A-1	09.1951	05.01.1968	
109	T.2-218	A-1	Oct 1951	13.10.1952	W/o 06.07.1952 at Hoyo de Manzanares, Madrid. Six killed
110	T.2-219, T.2B-219	A-1	10.1951	28.08.1970	
111	T.2-220, T.2B-220	A-1	11.1951	08.06.1955	W/o in landing accident 25.08.1954 at Valenzuela
112	T.2-221, T.2B-221	A-1	12.1951	04.02.1956	W/o 07.12.1955 at Cuatro Vientos in foggy weather
113	T.2-222, T.2B-222	A-1	12.1951	06.11.1957	
114	T.2-223, T.2B-223	A-1	02.1952	07.03.1956	W/o 15.11.1955
115	T.2-224	A-1 cv A-3	02.1952	02.08.1968	Suffered accident 27.09.1965 at el Aaiún
116	T.2-225, T.2B-225	A-1	03.1952	31.03.1969	
117	T.2-226, T.2B-226	A-1	03.1952	17.10.1955	W/o 10.03.1955 at Jerez. One killed
118	T.2-227	A-3	03.1952		W/o 12.01.1953 at Tétouan in landing accident
119	T.2-228, T.2B-228	A-1	04.1952	28.08.1970	
120	T.2-229, T.2B-229	A-1	04.1952	01.12.1965	Coll on ground with T.2B-230 at Alcantarilla, Murcia. Three killed
121	T.2-230, T.2B-230	A-1	04.1952	01.12.1965	Coll on ground with T.2B-229 at Alcantarilla, Murcia. Three killed
122	T.2-231, T.2B-231	A-1	04.1952	22.11.1974	
123	T.2-232, T.2B-232	A-1	05.1952	31.07.1971	
124	T.2-233, T.2B-233	A-1	05.1952	14.03.1963	W/o el Aaiún
125	T.2-234, T.2B-234	A-1	05.1952	12.06.1956	W/o Valencia
126	T.2-235, T.2B-235	A-1	06.1952	12.03.1969	W/o Taralejos
127	T.2-236, T.2B-236	A-1	06.1952	04.10.1958	
128	T.2-237, T.2B-237	A-1	06.1952	01.12.1972	
129	T.2-238, T.2B-238	A-1	08.1952	21.11.1966	
130	T.2-239, T.2B-239	A-1 cv A-3 cv A-1	07.1952	06.11.1959	W/o 23.06.1959
131	T.2-240, T.2B-240	A-3	07.1952	14.07.1971	
132	T.2-241, T.2B-241	A-1	07.1952	21.10.1972	
133	T.2-242	A-1	09.1952	12.02.1955	
134	T.2-243, T.2B-243	C	09.1952	10.06.1959	
135	T.2-244, T.2B-244	C	12.1952	17.11.1970	Preserved at NMUSAF, USA
136	T.2-245, T.2B-245	C	06.1953	20.02.1969	
137	T.2-246, T.2B-246	C	06.1953	08.02.1974	Preserved at Torrejón AB, Spain

138	T.2-247	?	06.1953	16.04.1956	W/o 01.11.1954 at Serranía de Ronda, Malaga. Eleven killed
139	T.2-248, T.2B-248	C	07.1953	20.02.1969	
140	T.2-249, T.2B-249	C cv A-1	07.1953	12.03.1970	W/o 06.12.1969 at Bir Enzarán, Sahara Desert
141	T.2-250, T.2B-250	A-1	07.1953	12.07.1972	
142	T.2B-251	A-1	07.1953	05.12.1957	W/o 08.05.1957 at Sidi Ifni. Four killed
143	T.2B-252	A-1	08.1953	02.12.1960	W/o 20.06.1960 at Manises
144	T.2B-253	A-1	11.1953	31.07.1959	
145	T.2B-254	A-1	08.1954	02.12.1972	Preserved at Museo del Aire, Cuatro Vientos, Spain
146	T.2B-255	A-1	08.1954	02.12.1972	Preserved at NASM, USA
147	T.2B-256	A-1 cv A-3	08.1954	31.08.1971	
148	T.2B-257	A-1	08.1954	30.11.1974	
149	T.2B-258	A-1	08.1954	16.06.1963	W/o 27.01.1962 at Edcheiría, Sahara Desert
150	T.2B-259	A-1	08.1954	07.02.1963	
151	T.2B-260	A-1, cv A-3	07.1954	24.02.1973	
152	T.2B-261	A-1	08.1954	10.06.1969	
153	T.2B-262	A-1	08.1954	01.06.1973	Preserved at Fantasy of Flight, USA
154	T.2B-263	A-1	08.1954	10.06.1969	
155	T.2B-264	A-1	08.1954	28.08.1970	
156	T.2B-265	A-1	12.1954	02.12.1972	
157	T.2B-266	A-1	10.1954	30.08.1966	
158	T.2B-267	A-1	12.1954	23.01.1970	
159	T.2B-268	A-1	12.1954	20.12.1955	W/o 27.06.1955 at Tan Tan, Sahara Desert. One killed
160	T.2B-269	A-1	12.1954	12.02.1971	
161	T.2B-270	A-1	12.1954	04.10.1955	
162	T.2B-271	A-1	12.1954	17.10.1957	W/o 29.05.1957 at Aras de Alpuente, Valencia. Eight killed
163	T.2B-272	A-1	12.1954	02.12.1972	Preserved at RAF Museum, Cosford. Marked as 'G-AFAP'
164	T.2B-273	A-1	12.1954	21.10.1972	
165	T.2B-274	A-1	12.1954	02.12.1972	
166	T.2B-275	A-1	12.1954	01.06.1973	
167	T.2B-276	A-1	02.1955	13.10.1970	
168	T.2B-277	A-1	02.1955	17.11.1961	
169	T.2B-278	A-1	02.1955	11.10.1958	W/o 14.07.1955 at Armilla Granadá
170	T.2B-279	A-1	05.1955	12.08.1959	W/o 18.06.1959 at Alcantarilla

Endnotes

Introduction

1 Brown, Eric, 'Iron Annie from Dessau', *Air International*, October 1975, p. 180.

Chapter 1

1 Pletschacher, Peter, *Lufthansa Junkers Ju 52* (1990), p. 38f.
2 Grosz, Peter M., *Junkers D.I*, Windsock Datafile 33, p. 7.

Chapter 2

1 In 1941, the Petlyakov Pe-2 twin-engined bomber was entering production at GAZ No. 7. However, with German forces on Moscow's doorstep, the factory was evacuated to Kazan. The Limhamn facility in Sweden closed down in 1935.
2 The Junkers G 38 design was sold to Japan, where a few were constructed for use as bombers. The second and last Dessau-built G 38 were eventually transferred to the Luftwaffe, being destroyed at Tatoi, Greece, in May 1941 during an RAF raid.
3 Various dates for the first flight of the Ju 52 have been published, including 13 October.
4 Ishoven, Armand van, 'Udet (BFW) U-12 Flamingo Variants', *Aircraft in Profile* 257, p. 102.

Chapter 4

1 On 30 June 1933, the name changed from *Deutsche Luft Hansa AG* to Deutsche Lufthansa AG. For the sake of simplicity, the name Lufthansa will be used throughout this book.
2 Pletschacher, Peter, *Lufthansa Junkers Ju 52*, p. 148f.
3 An incomplete list of Ju 52/3ms lost in accidents is available on aviation-safety.net. The first German-registered Ju 52/3m to be written-off in an accident appears to have been D-AVAN on 6 November 1934. At the time, the aeroplane was operated by *Deutsche Reichsbahn-Gesellschaft.*
4 Quoted in Pletschascher, Peter, *Lufthansa Junkers Ju 52*, p. 67.
5 'Over the Roof of the World', *Shell Aviation News*, January 1938 pp. 8–10.

Chapter 5

1 Hagedorn, Dan, *The Trek of Aconcagua*, p. 225.

Chapter 6

1 Hooton, E. R., *Phoenix Triumphant*, p. 103.
2 The Gotha 244 was essentially a powered variant of the Go 242 assault glider. Apart from the new production Go 244s, another 133 were converted from Go 242s. The six-engined Messerschmitt Me 323 was the largest transport in Luftwaffe service, being able to carry 120 fully equipped troops or sixty stretcher patients.

Chapter 7

1 The exact number of Ju 52/3ms produced in France between 1942 and 1944 is unclear. However, the first AAC.1 delivered in December 1944 carried a dual construction number, 539/001. This may indicate that 538 Ju 52/3ms were built during the German occupation. Other sources indicates figures as low as 516 and as high as 602 Ju 52/3ms built by Amiot.
2 Hooton, E. R., *Phoenix Triumphant*, p. 122
3 Kightey, James, 'Air Gunner Junkers Ju 52', *Aeroplane Monthly*, April 2008, p. 27
4 Quoted in Sadler, John, *Operation Mercury*, p. 69.

Chapter 9

1 Hooton, E. R., *Phoenix Triumphant*, p. 62.
2 Kightey, James, 'Air Gunner Junkers Ju 52', *Aeroplane Monthly*, April 2008, pp. 26–27. The numerical suffix indicates the weight of the bomb measured in kilogrammes.
3 Hooton, E. R., *Phoenix Triumphant*, p. 115.
4 The abbreviation KGr b z b V, will be used hereafter.

Chapter 10

1 Ketley, Barry, *Fledgling Eagles*, p. 62f.

Chapter 12

1 Shores *et al.* (1987), p. 108.
2 Shores (2012), p. 378.
3 Bergström, Christer, *The Air War Over the Eastern Front: Resurgence January–June 1942*, p. 89.

Chapter 13

1 Bergström, Christer, *Stalingrad: The Air Battle: 1942 through January 1943*, p. 104.
2 Bergström, Christer, *Stalingrad: The Air Battle: 1942 through January 1943*, p. 123.

Chapter 14

1 The Ju 52/3m was easily recognisable by Allied fighter pilots, with a contemporary USAAF film on aircraft recognition stating that 'You can't miss it, it's the only three-motored plane in the German Air Force'.

Chapter 15

1 Pegg, Martin, *Transporter Vol. 2*, p. 166. The contents of these supply containers had been donated by *Reichsführer* Heinrich Himmler in remembrance of the fact that the Nazi regime had seized power on 30 January 1933. A formal complaint against this unwanted delivery was made by the garrison.
2 Pegg, Martin, *Transporter Vol. 2*, p. 163.

Chapter 16

1 Lee, Russell, 'Dissecting an Air Force', *AAHS Journal*, Spring 1999, p. 47. Interestingly, the Junkers company library at Dessau, containing upwards of 10,000 volumes, was surrendered to US forces.

Chapter 17

1 Darbyshire, Don, 'Booby's 'Aunties'', *Air Enthusiast*, Issue 101, p. 11.
2 Andersson, Lennart, 'Chinese 'Junks'', p. 7. For a detailed history on Eurasia, see Eurasia Aviation Corporation by Morgenstern, K., and Plath, D.
3 Wegg, John, *Finnair: The Art of Flying Since 1923*, (1983), p. 55.
4 Wegg, John, *Finnair: The Art of Flying Since 1923*, (1983), p. 67f.
5 Sims, Charles, *Camera in the Sky*, p. 191.
6 The former British Airways Ju 52/3m was subsequently registered as D-AEHB, being noted at Bromma airport on 22 September 1941.
7 Seabrook-Smith, Harold, 'An Anglo-Nordic Saga', *Aeroplane Monthly*, December 1974 p. 1041.
8 Norrman (1994), p. 20.
9 Whittier, Bob, 'The Fascinating Junkers Ju-52', *Sport Aviation*, October 1972, p.p 55–56.
10 Lennart Andersson's book *Aeroflot Origins* (Apali Oy 2009) contains much additional details on Soviet Ju 52/3m operations.
11 Andersson, Lennart, *Aeroflot Origins*, (2009), p. 12ff.

Chapter 18

1 Windrow, Martin, (2005), p. 524.
2 The complete history of this particular Ju 52/3m can be found in Hagedorn, Dan., 'The Trek of the Aconcagua', published in the *AAHS Journal*, Fall 1992.

Bibliography

Archives

ABA-SAS Historical Collection
Arlanda Flygsamlingar (Arlanda Civil Aviation Collection)
Riksarkivet (Swedish National Archives)
Krigsarkivet (Swedish National War Archives)

Books

Andersson, L., *A History of Chinese Aviation*, (Taiwan: AHS of ROC, 2008); *Aeroflot Origins: Red Stars 6*, (Finland: Apali Oy, 2009)

Andersson, L., Endres, G., Mulder, R. J. M., & Ott, G., *Junkers F 13: The World's First All-Metal Airliner*, (United Kingdom: EAM Books EEIG, 2012)

Bernád, D., *Rumanian Air Force, The Prime Decade, 1937–1947*, (USA: Squadron/Signal Publications, 1999)

Billig, D., & Meyer, M., *Flugzeuge der DDR. 1. Band: Bis 1962*, (Germany: 2002)

Butler, P., *War Prizes*, (Leicester: Midland Counties Publications, 1994)

Dancey, P., & Vajda, F., *German Aircraft Industry & Production, 1933–1945*, (Shrewsbury: Airlife Publishing Ltd, 1998)

Davies, R. E. G., *Lufthansa: An Airline and its Aircraft*, (New York: Orion Books, 1991)

Erfurth, H., *Flugzeug-Legende JU 52*, (Munich: GeraMond Verland, 2013)

Green, W., *The Warplanes of the Third Reich*, 4th Edition, (Great Britain: Doubleday and Company, Inc., 1979)

Gunston, B., *World Encyclopedia of Aircraft Manufacturers*, (United Kingdom: Patrick Stevens Limited, 1993)

Hagby, K., *Fra Nielsen & Winther til Boeing 747*, (Drammen: 1998)

Hagedorn, D., *Latin American Air Wars and Aircraft 1912-1969*

Hooton, E. R., *Phoenix Triumphant: The Rise and Rise of the Luftwaffe*, (London: Arms & Armour Press, 1994)

Hoppe, H., *Junkers Ju 52/3m*, (Erlangen: AirDOC, 2004)

Kay, A. L., *Junkers Aircraft & Engines 1913–1945*, (London: Putnam Aeronautical Books, 2004)

Lybye, K., *Det danske Luftfartsselskab gennem 25 Aar*, (Copenhagen: 1943)

Miertsch, W., *Vom Original zum Modell: Junkers Ju 52*, Zivil Versionen (Civil Version), (Bonn: Bernhard & Graefe Verlag, 1998)

Morgenstern, K., & Plath, D., *Eurasia Aviation Corporation*, (Munich: GeraMond Verland, 2006)

Pavón, L. G., *El Junkers Ju 52/3m CASA C-352*, (Madrid: 2014)

Pegg, M., *Luftwaffe Transport Units 1943-1945*

Piekalkiewicz, J., *Die Ju 52 im Zweiten Weltkrieg*, (Stuttgart: Motorbuch Verlag, 1977)

Pletschacher, P., *Lufthansa Junkers Ju 52: The Story of the Old 'Aunty Ju'*, (Landshut: Aviatic Verlag, 1990)

Rajich, J., & Sehnal, J., *Slovakian Airmen 1939-1945*, (Kutná Hora, 1991)

Rosch, B., *Luftwaffe Support Units*, (Hersham: Ian Allan Publishing, 2009)

Shores, C., Cull, B., and Malizia, N., *Air War for Yugoslavia, Greece and Crete 1940–41*, (London: Grub Street, 1987)

Shores, C., Massimellio, G., & Guest, R., *A History of the Mediterranean Air War 1940-1945: Volume One: North Africa June 1940–January 1942*, (London: Grub Street, 2012)

Sims, C. A., *Camera in the Sky*, (London: Temple Press Ltd, 1958)

Stafrace, C., *Arab Air Forces*, Aircraft Special series (6066), (Carrollton: Squadron/Signal Publications, 1994)

Stapfer, H., Mau, H., & Punka, G., *Junkers Ju 52 in action*, (Carrollton: Squadron/Signal Publications, 2002)
Urech, J., *The Aircraft of the Swiss Air Force since 1914*, (Stäfa: Verlag Th. Gut + Co, 1975)
Wegg, J., *Finnair: The Art of Flying Since 1923*, (Finnair Printing Department, 1983)
Windrow, M., *The Last Valley*, (London: Cassell Military Paperbacks, 2005)

Magazines and Articles

Aeroplane 1936–1939, 1973–2014, various issues
Andersson, L., 'Chinese 'Junks'', *Air Enthusiast*, No. 55, Autumn 1994)
Becker, D,. 'Tante Ju in South Africa', *Aeroplane Monthly*, March 1988, pp. 152–155
Cowin, H., 'The Junkers Monoplanes', *Aircraft in Profile*, 187.
Flugzeug Classic, various issues
Grant, R. S., 'Canada's Corrugated Classic', *Air Enthusiast* 30, pp. 70–76
Gwynn-Jones, T., 'Flying For Gold', *Aeroplane Monthly* July 1991, pp. 420–423
Hagedorn, D., 'The Trek of the Aconcagua', *AAHS Journal*, Fall 1992, pp. 223–229
Jakob, J. A., 'Bolivian 'Tin'', *Air Enthusiast* 72, pp. 2–7
Jane's All the World's Aircraft, various editions 1930–1947
Jet & Prop, various issues
Lee, R., 'Dissecting an Air Force', *AAHS Journal*, Spring 1999, pp. 46–55
Mladenov, A., 'Balkan Bulgarian Airlines', *Air International*, April 2000, pp. 229–235
Norrman, I., 'Prinsen av Wales flyger norrut', *Svensk Flyghistorisk Tidskrift*, No. 1, 1994, pp. 18–20
Olsson, L. E., 'Tant Ju 50 år: snäll, villig och fortfarande flygvärdig', *MACH* Issue 11, pp. 44–47
Rauch, G. von, 'The Green Hell Air War', *Air Enthusiast*, Quarterly 2, pp. 207–213; 'A South American Air War… The Leticia Conflict', *Air Enthusiast* 26, pp. 1–8
Seabrook-Smith, H., 'An Anglo-Nordic saga', *Aeroplane Monthly*, December 1974, pp. 1041–1045
Shell Aviation News 1934–1939
Shores, C., 'The Fight For Iraq: 1941', *Air Enthusiast* 10, pp. 26–39
Stroud, J., 'Wings of Peace', *Aeroplane Monthly*, February 1985, pp. 97–101
Whittier, B., 'The Fascinating Junkers Ju-52', *Sport Aviation*, October 1972, pp. 55–56
Widfeldt, B., 'Tante Ju', *I luften Flygets Årsbok*, 1983, pp. 68–85

Internet

aviation-safety.net
www.aeroflight.co.uk
www.airhistory.org.uk
www.baha.be
www.junkers.de

Index